Drones and Targeted Killing in the Middle East and Africa

Drones and Targeted Killing in the Middle East and Africa

An Appraisal of American Counterterrorism Policies

Christine Sixta Rinehart

LEXINGTON BOOKS
Lanham • Boulder • New York • London

Published by Lexington Books
An imprint of The Rowman & Littlefield Publishing Group, Inc.
4501 Forbes Boulevard, Suite 200, Lanham, Maryland 20706
www.rowman.com

Unit A, Whitacre Mews, 26-34 Stannary Street, London SE11 4AB

British Library Cataloguing in Publication Information Available

Library of Congress Cataloging-in-Publication Data Available
ISBN 978-1-4985-2647-0 (cloth : alk. paper)
ISBN 978-1-4985-2648-7 (electronic)

∞™ The paper used in this publication meets the minimum requirements of American National Standard for Information Sciences—Permanence of Paper for Printed Library Materials, ANSI/NISO Z39.48-1992.

Printed in the United States of America

Contents

Acknowledgments

This book was inspired foremost by a friend and colleague President Nayef Samhat of Wofford College. During my time as a Visiting Assistant Professor in the Government Department in fall 2013, I was asked by Nayef if I would be interested in speaking to his friend, Professor Amos Guiora from the S.J. Quinney College of Law at the University of Utah, concerning his work on counterterrorism. I jumped at the chance to meet Professor Guiora and hear him speak, in addition to being a guest lecturer in my International Terrorism class. Although Professor Guiora's work is on the legal criteria and justification for targeted killing, I was immediately fascinated by the subject matter and asked Amos a million questions. What followed were two great friendships. Thank you to both of you for your time and camaraderie. This book would never have been written without that initial introduction.

The idea for this book began as a conference paper for the Association for the Study of the Middle East and Africa (ASMEA) Annual Meeting. I realized that one paper could not possibly examine this entire subject and so I began writing the book and luckily, the data materialized from numerous sources. For those of you who have spent numerous hours compiling data like Chris Woods, thank you for your work!

I also wish to thank Colonel Adam Oler for his invitation to Holloman Air Force Base so that I could meet our amazing pilots and Air Force personnel. I had strong reservations writing this book as I did not want to criticize our military or the hard work they put in to defending the United States. It is with all of you in mind that I state that this book is an evaluation of the foreign policy that the Executive Branch has chosen to pursue including George W. Bush and Barack H. Obama. I have tremendous gratitude and respect for the United States' Air Force and our military.

I also wish to thank the anonymous reviewer who provided several useful suggestions and comments. I also have several friends and colleagues to thank including my long-time close friend and former Chairmen Dr. Kevin Lasher at Francis Marion University for his loyal friendship and eternal belief in me. In addition, I am grateful for my former Academic Dean at USC Union Dr. Stephen Lowe who is also a first-rate dean and colleague. You are always looking out for us. I also wish to thank my husband Todd who cherishes me and holds me to the strictest standards, always pushing me to be better. You are everything to me and my best friend.

I also want to thank the University of South Carolina for their financial support during the time I wrote this book. This book was funded by a Research Initiative for Summer Engagement (RISE) grant from the University of South Carolina's Office of the Vice President for Research. I am truly grateful for the financial assistance and support from the University of South Carolina. I am honored to be faculty at this Carnegie One Research Institution. Go Cocks!

Introduction

The first United States' drone strike occurred on November 3, 2002, when a US Predator drone fired a Hellfire missile at a car traveling through the Mar'ib province of Yemen, killing six people inside the car; one of the dead was an American citizen Kamal Derwish. Since this first American drone strike, American drone usage has increased tremendously. According to the author's calculations using data cited in Table I.3: *Drone Strike Data*, during the Presidency of George W. Bush, about two drone strikes were ordered per day in the Middle East. From his first day in office until the end of 2015, approximately ten drone strikes per day have been ordered in the Middle East and Africa during Obama's presidency. During George W. Bush's presidency around 5,000 drone strikes were ordered and under President Obama around 25,000 drone strikes have been ordered until the end of 2015. Bergen and Rowland found that under President Bush, about 33 percent of drone strikes in Pakistan killed a leader compared to less than 13 percent under Obama.[1] Bush targeted al-Qa'ida members in 25 percent of drone attacks while Obama targets Taliban members in approximately 40 percent of all drone attacks.[2] Although the United States' military has embraced drones as the highly anticipated, low casualty, warfare weapon of the future, the effects of drone warfare are questionable.

The United States has repeatedly used drones to kill suspected terrorists in Afghanistan, Iraq, Libya, Pakistan, Somalia, Syria, and Yemen in an effort to decrease terrorism and the vitality of terrorist groups. Targeted killing through the use of drones has become a foreign policy weapon to keep the United States safe from further terrorist attacks. However, it is suspected that targeted killings through the use of drones have actually led to an increase in terrorist group recruitment, terrorist attacks, and empathy for the terrorist group from the local population in addition to several other unwanted repercussions. The two part research question this book attempts to answer is, "What is the effect of drone targeted killing on Afghanistan, Iraq, Libya,

Pakistan, Somalia, Syria, and Yemen? And is it a successful method in the War on Terror?"

The first part of the research questions seeks an answer as to how the countries have actually fared or responded to drone warfare and targeted killing in their countries. For example, it has been found in Pakistan that targeted killing in rural areas where al-Qa'ida trains has caused al-Qa'ida members to move to cities. This has caused an increase in crime and radicalization in the cities. There is also a political party in Pakistan devoted to the discontinuation of American drone use in Pakistan called the Tehreek-e-Insaf (PTI).[3,4] Each chapter in this book will examine one particular country and how drone warfare has affected that country whether it be the government, citizens, crime rates, etc. The primary goal of this book is to see whether targeted killing is effective as a foreign policy tool.

The second component of the research question examines whether targeted killing has been effective in the decrease of terrorism in each country and in the United States. The purpose of the War on Terrorism was foremost to protect the United States' homeland and to decrease terrorism and terrorist groups throughout the world. George W. Bush stated, "Our war on terror begins with al-Qaeda, but it does not end there. It will not end until every terrorist group of global reach has been found, stopped and defeated."[5] Each chapter on each country will analyze whether terrorism has really decreased in that country since targeted killing began. The conclusion will assess the results of targeted killing in the Middle East and Africa and whether it has decreased terrorism overall. The conclusion will also examine whether targeted killing has decreased terrorism in the United States and will look at some preliminary costs of drone warfare.

Preliminary research for this book found that Pakistan had a correlation between targeted killing and an increase in suicide bombings within Pakistan, which is counterproductive for Pakistani national security and safety. In reference to the United States, an argument can be made that as long as the American homeland is not getting attacked, who cares what happens to the Pakistanis? The response to this question is that American values and culture do not allow for a lack of apathy for other countries and citizens in the worldwide War on Terror. If Americans just do not care about what happens to citizens in other countries on behalf of their own government, then why should citizens in other countries ever help Americans? The ends do not justify the means, particularly if reciprocity is needed somewhere in the relationship. In addition, apathy is not conducive to creating a good alliance with the Pakistanis when the United States is trying to destroy terrorist elements through other methods such as freezing assets or raiding an al-Qa'ida complex. Strong diplomatic relations with a country like Pakistan are a necessity in the War on Terror. In addition, targeted killing is supposed to decrease terrorism in general, not only in the United States.

Since this is a relatively new area of research, this book is literally the first crack at summation and analysis of drone warfare in the Middle East and Africa. Unfortunately, someone must always be first, so there will undoubtedly be some mistakes and some room for improvement. All mistakes are of course my own. All of the data used in this book are taken from think tanks that are independently gathering data, not the United States' Government. The disparities between what the United States Government actually rarely reports and what occurs are tremendous. One can be assured that this question of targeted killing effectiveness will continue to attract more research and analysis as time marches onward, particularly since future presidential administrations, whether under Hillary Clinton or Donald Trump are not opposed to the policy.

This introductory chapter will continue with a brief history of drone warfare. The next part of the introduction will examine current literature on whether targeted killing is effective or ineffective. The next section will include information on quantitative data operationalization, and terminology including unmanned aerial vehicles (UAV), and actors (targeters and targets). The introductory chapter will then explain what it means for a country like the United States to effectively decrease terrorism in the War on Terror or how is effective counterterrorism operationalized in the book? The last part of this chapter will provide an overview of the book and a summary of each chapter.

A BRIEF HISTORY OF DRONE WARFARE

Although there are more simplistic earlier examples of drones, Nikola Tesla created a radio-guided electronic torpedo in 1897 and a radio-controlled motorboat in Madison Square Garden in 1898. Charles Kettering then introduced the Kettering Bug or an aerial torpedo in World War I. World War II led to greater advances with both Germany and the United States developing their own remote-controlled planes. Joe Kennedy, the eldest brother of John F. Kennedy, actually died as a result of a naval drone explosion gone wrong in World War II.

Drones continued to develop throughout the Cold War and were even extensively used in the first Gulf War in Iraq. The United States bought the Pioneer from Israel and proceeded to use it in Iraq. Bill Clinton even used drones to spy on Osama bin Laden in Afghanistan, even though drones did not yet have destruction capabilities. After September 22, 2001, the United States greatly increased its manufacturing of drones, drone weapons capabilities, and drone bases around the world. By the end of October 2001, the United States had created a framework for targeted killings using drones as the CIA had already been experimenting with the Predator in Afghanistan and Pakistan concerning surveillance operations.[6] "The United States has at least sixteen drone operating and training sites domestically along with another twelve American drone bases or support facilities abroad in Djibouti, Ethiopia, Kuwait, Niger, Oman, Saudi Arabia, Seychelles, Qatar, and the UAE."[7]

Drone or UAV[8] technology has greatly changed the world of crime and thus what follows, security. In their thoughtful and well-researched book *The Future of Violence*, Benjamin Wittes and Gabriella Blum examine how UAV technology has changed the face of national and international security. They argue the following.

1. Modern technology enables individuals to wield the destructive power of states.
2. Individuals, including you personally, can potentially be attacked with impunity from anywhere in the world.
3. Technology makes less relevant many of the traditional concepts around which our laws and political organization for security have evolved. National borders, jurisdictional boundaries, citizenship, and the distinctions between national and international between act of war and crime, and between state and private action all offer divides less sharp then they used to.
4. Our nation—and every nation—can face attack through channels controlled and operated not by governments but by the private sector and by means against which governments lack the ability to defend, making private actors pivotal to defense.[9]

In addition to the security environment, drones have tremendously changed the way that the United States conducts itself in the War on Terror. The United States no longer needs to put "boots on the ground" to fight a foreign enemy. Drones are capable of doing that for all the military branches and less lives and equipment are lost.[10] Targets can be destroyed by pilots flying from thousands of miles away. The next part of this chapter will examine existing literature concerning the effectiveness of targeted killing.

WHAT DOES THE LITERATURE SAY?

Several studies have been completed concerning the efficacy of drone strikes on terrorist organizations, yet there is still no consensus. While it is true that drone strikes kill terrorists or cause them to flee, it is also true that local populations are incensed and radicalized by American drone strikes. The following literature review provides the most current consensus for both the effectiveness and ineffectiveness of drone strikes.

Drone Targeted Killings are Effective

Some scholarship has been published concerning the effectiveness of drone strikes in combating terrorism. Professor Daniel Byman[11] asserts that drones work because: they have killed several thousand al-Qa'ida members, undermine communicative technology like cellphones, have a low ratio of civilian to militant deaths, and are supported by foreign governments. Professor Wilner gives six reasons why he believes targeted killings work, not necessarily using drones. (1) The first is the elimination of the threat. (2) The second is that targeted killing forces a leader to look after their own safety as opposed to any activities conducted by their terrorist group. (3) Thirdly, targeted killings deprofessionalize an organization, thus leaving them not to do anything besides staying alive. (4) The fourth reason that targeted killing is successful is that it leads to disorganization and power struggles within the organization. (5) The fifth and final reason targeted killings work is that they lead to low moral within the organization. After a successful series of targeted killings, terrorist members will feel despair.[12]

In a study using data from January 2007 to September 2011, RAND analyst Patrick B. Johnston and Stanford University scholar Anoop K. Sarbati found that drone strikes were effective in decreasing terrorist attacks in northwest Pakistan and eastern Afghanistan (FATA). The authors found that drone strikes were also effective in decreasing the lethality of terrorist attacks. The drone strikes were most effective during the week that the drone strikes occurred but had little long-term results in decreasing terrorist attacks.[13]

In summation concerning the effectiveness of targeted killings, the jury is still out and the academics and policy makers who support drone strikes are less in number than those who can see the problems with drone strikes. Strictly speaking, even the most pro-drone advocates like Byman caution the United States against using drone strikes as the main counterterrorism tactic in the War on Terrorism. The lack of humanity in killing a human being thousands of miles away on a computer screen is dehumanizing and risky for many reasons. In the end, the cautious proponents are correct, particularly under the Obama Administration. Targeted killing and drone warfare have become the sole tactic in the War on Terror and the overreliance of the United States on this method is readily apparent. The next section will describe the numerous problems with drone warfare and targeted killing.

Drone Targeted Killings are Not Effective

There are several problems with drone targeted killings:

(1) The first of these problems is that innocents are often killed, including American service people,[14] and collateral damage occurs. Whether it is mechanical error, human error, or any other reason, mistakes are made. Professor Audrey Cronin[15] warns in a *Foreign Affairs* article that drones "may be creating sworn enemies out of a sea of local insurgents" and that "It would be a mistake to embrace killer drones as the centerpiece of U.S. counterterrorism." Noor Behram, who is known for photographing the effects of drone

strikes in Pakistan, has stated "For every ten to fifteen people killed, maybe they [American Military] get one militant."[16] People believe in the accuracy of American weaponry and mistakes are taken for purposeful actions.[17] Since many Muslims in the Middle East and Africa believe that the Americans are crusaders or that they are trying to control the world, the best of intentions create the worst perceptions when mistakes are made. In addition, since bodies are incinerated when drone strikes occur, Muslims (who are typically the target population) cannot perform proper burials because there is nothing left of the body.[18] As a side note, after the drone strikes occur, al-Qa'ida and the Taliban have been known to cordon off the strike area. The drones will then return to the area to kill[19] the Taliban or al-Qa'ida members who have come to investigate, creating a drone funeral. This concept is called the "double tap" and was most likely invented by the Israelis "meaning that first responders to an area already hit, lost their own lives in the second round of bombing."[20] Although in recent months, the terrorist groups have got smarter in not going to the scene of the strikes, there are many instances throughout this book, where civilians who assist at the scene of the drone strikes become the victims of the second drone strikes.

In addition, there are also serious problems determining who exactly is a "terrorist." There is no universal definition of terrorism that has been agreed upon by any agency and therefore no concrete consensus of what constitutes a terrorist. In this book, the author is using the following definition of terrorism, which I used in my first book. In previous research, the author has distinguished between the two terms: terrorist and revolutionary. The definition of terrorism used in this book, which is differentiated from revolution, is as follows:[21] Terrorist groups are nongovernmental[22] entities that intentionally attack civilian populations who are noncombatants within their own struggle for power. Specifically, noncombatants refer to people who are not involved in any way with the terrorist's struggle and cannot really change the outcome of the terrorist organization's situation. For example, the president of a country may be a legitimate revolutionary target but a postal worker or bridge builder is not. Purposefully attacking people who hold political and military power is an act of war, not an act of terrorism. These people who hold political and military power would include off-duty soldiers or reserve soldiers and also public officials. Revolutionaries who use terrorist tactics (like the White Terror in the French Revolution where thousands were killed) become terrorists. One who uses terrorist tactics is a terrorist. Terrorists may subscribe to revolutionary ideology, but the killing or terrorizing of noncombatants is what defines a terrorist as a terrorist, and differentiates the terrorist from a revolutionary. As a side note, this definition of terrorism does not include state-sponsored terrorism, which the author believes is an entirely separate topic and is somewhat of a misnomer.[23]

Besides the problem with a universal definition of terrorism, what physically constitutes a terrorist to the American federal government is liberal and comprehensive, at least according to the think tanks who are tracking civilian deaths. Anyone who has any affiliation with a terrorist group, no matter whether coerced or willingly, can be considered a target. This apparently includes wives, children, truck drivers, gardeners, etc. or anyone who briefly comes into contact with the terrorist. Although there may be a lot of accidents or collateral damage, the United States' Government does not appear to discriminate when targeting some of these people.

Prior to the publication of *Executive Order—United States Policy on Pre- and Post-Strike Measures to Address Civilian Casualties in U.S. Operations Involving the Use of Force*[24] on July 1, 2016, the White House announced that it would give casualty and strike numbers in the Executive Order. According to *The Situation Report, The National Security Daily Brief* from *Foreign Policy*,[25] these numbers were expected to be 100 civilian deaths in Pakistan, Libya, and Somalia with around 500 drone strikes just in 2016. However, these numbers never materialized in the Executive Order and instead the White House stated that it would report on the numbers every year from then on. The numerous sources used throughout this book strongly dispute these White House numbers for 2016. By President Obama's own admission, numbers for Afghanistan, Iraq, and Syria were not to be included in the expected report. Afghanistan alone has the highest number of civilian deaths as will be discussed in Chapter 2.

(2) The second problem with drone targeted killings is the lack of legal procedure that accompanies them. An individual is found guilty of committing an act of terrorism, yet evidence is never presented and there is never a trial. The United States' Government has been given blanket authority to kill whomever they deem a "terrorist" wherever they want. Most of the people targeted are too poor and too insignificant to ever fight back. President George W. Bush first began to use drones for assassinations. National Security Advisor Condoleezza Rice at the time stated, "The president has given broad authority to U.S. officials in a variety of circumstances to do what they need to do to protect the country. We're in a new kind of war, and we've made very clear that it is important that this new kind of war be fought on different battlefields. It's broad authority."[26] Jeffrey Smith, the CIA's former general counsel added "To the extent that you do more and more of this, it begins to look like it is policy. [If used regularly, such attacks would] suggest that it's acceptable behavior to assassinate people. . . . Assassination as a norm of international conduct exposes American leaders and Americans overseas."[27]

In the Obama Administration, decisions are often made by a few individuals in the military and the president on "Terror Tuesdays."[28] Little procedure exists. Sounding somewhat mythical, the president and 100 members of the national security agencies gather round a table and decide who should be next. Questions such as "should an al-Qa'ida facilitator die?" are mixed in with "does opening a gate for a truck filled with explosives make someone an al-Qa'ida facilitator?" President Obama admits that a certain amount of mistakes are going to be made. He is comfortable with the use of force and is confident in his decisions to use force except when it comes to putting "boots on the ground." However, he has no reservations when secretly killing people with drones, although the mass media has made little attempt to expose Obama's secret drone war.[29] Most Americans are oblivious as to what is really occurring with drones.

President Obama's Department of Justice *White Paper*[30] allows the military to kill American citizens outside of the United States for suspected terrorist actions despite the *Bill of Rights* that gives a citizen the right to a trial by jury (due process). It is also implied that American citizens who are "considered" an imminent threat in the United States will also be killed. Sixteen-year-old Abdurahman Anwar al-Awlaki was on the receiving end of a drone in Yemen and never received any trial, nor was he charged with any crime despite the fact that he was an American citizen.[31,32]

The egregious error of killing an American sixteen-year-old boy, Abdurahman, along with his cousins appears unconscionable. Although this book is not about the rights of American citizens, the following lengthy quote from Jeremy Scahill about Abdurahman must be included in this introduction.

> The young teenager sat outside with his cousins as they gathered for a barbeque. He wore his hair long and messy. His mother and grandparents had repeatedly urged him to cut it. But the boy believed it had become his trademark look and he liked it. A few weeks earlier, he had run away from home, but not in some act of teenage rebellion. He was on a mission. In the note he left for his mother before he snuck out the kitchen window as the sun was just rising and headed to the bus station, he admitted that he had taken money from her purse-$40-for bus fare, and for that he apologized. He explained his mission and begged for forgiveness. He said he would be home soon.
>
> The boy was the eldest child in his family. Not just in his immediate family, which consisted of his parents and his three siblings, but in the large house they shared with his aunts and uncles and cousins and two of his grandparents. He was his grandmother's favorite. When guests visited, he would bring them tea and sweets. When they left, he would clean up after them. Once, his grandmother twisted her ankle and went to the hospital for treatment. When she limped out of the treatment room, the boy was there to greet her and make sure she got home safely. "You are a gentle boy," his grandmother always told him. "Don't ever change."
>
> The boy's mission was simple: he wanted to find his father. He hadn't seen him in years and he was afraid that if he didn't find him, he would be left

> only with blurred memories: of his father teaching him to fish; to ride a horse; surprising him with an abundance of gifts on his birthday; taking him and his siblings to the beach or to the candy shop.
>
> Finding his father would not be easy. He was a wanted man. There was a bounty on his head and he had narrowly escaped death more than a dozen times. That powerful forces in multiple countries wanted his father dead did not deter the boy. He was tired of seeing the videos of his father that painted him as a terrorist and an evil figure. He just knew him as his dad, and he wanted at least one last moment with him. But it didn't work out that way.
>
> Three weeks after he climbed out the kitchen window, the boy was outdoors with his cousins—teenagers like him—laying a picnic for dinner beneath the stars. It was then he would have heard the drones approaching, followed by the whiz of the missiles. It was a direct hit. The boy and his cousins were blown to pieces. All that remained of the boy was the back of his head, his flowing hair still clinging to it. The boy had turned sixteen years old a few weeks earlier and now he had been killed by his own government. He was the third U.S. citizen to be killed in operations authorized by the president in two weeks. The first was his father.[33]

When questioned about the American Government's assassination of American people, former United States' Attorney General Eric Holder stated in March 2012:

> Let me be clear: An operation using lethal force in a foreign country, targeted against a U.S. citizen who is a senior operational leader of al Qaeda or associated forces, and who is actively engaged in planning to kill Americans, would be lawful at least in the following circumstances: First, the U.S. government has determine . . . that the individual poses an imminent threat of violent attack against the United States; second, capture is not feasible, and third the operation would be conducted in a manner consistent with applicable law of war principles.[34]

Regardless of the laws and justifications that the Obama Administration uses, the targeted killing of American citizens violates the *Bill of Rights* and the United States *Constitution* for the following reasons: It is difficult to believe that both Abdurahman Anwar al-Awlaki and the other American men that have been targeted by the Obama Administration were an imminent threat to the United States. It is understandable that an American with a bomb strapped on his back heading to a shopping mall is an imminent threat. The justification of the deaths of these American men was never given. If American citizens can be found by drones, why can they not be extradited to the United States for a trial, whether they are arrested by a foreign military or an American military? These men were killed in another country. It is not the job of the author to justify why killing American citizens is legal;

indeed concerning American laws, it is the job of the American Government to justify their actions. To date, these deaths have not been explained by the American Government.

On a semi-brighter note, President Obama stated on May 27, 2013, that the use of drones and targeted killing needs some oversight and structure, even asking Congress for advice. Lawyers were asked to present arguments on both sides of the debate but in the end the President's team did not want to give up any presidential power. Nothing has come out of these statements, over three years later.[35]

San Francisco Law Professor, Jeffrey S. Brand, and Utah Law Professor, Amos Guiora, have suggested the idea of a drone court that would be part of the judicial system, operating with judges confirmed by the Senate. The purpose of this drone court would be to prevent the president from unilaterally making decisions concerning targeted killings. The court would be independent from the executive branch and would consist of twenty-four Article III justices, twelve from the district court and twelve from the court of appeals chosen at random. Guiora and Brand provide a rationale and blueprint for the drone court in their coauthored chapter. The drone court is an innovative recommendation to solve the problem of the unitary executive and ensure due process.[36] However, ideas such as these have drawbacks and possible solutions will be discussed in the Conclusion.

(3) The third problem with drone targeted killings is that it is dehumanizing. Pilots are often flying these drones from the United States' military bases, their cockpits located in tractor trailers, without ever placing a foot in the war zone.[37] Pilots are able to go to work at a New Mexico or Nevada Air Force base, eat lunch at home, and kill terrorists in Yemen or Iraq all in a day's work. Drone warfare appears as a video game at an American military base. Although the pilots and Air Force refuse to equate the technology to a video game, this analogy is correct. Drone pilot Lt. Colonel Matt J. Martin talks about his training in the Predator Formal Training Unit (FTU) at Creech Air Force Base, Indian Springs, Nevada in his memoir *Predator*. He states,

> I was among the first generation of soldiers working with robots to wage war. The ability to kill people from such great distances, playing God, widened the gap between the reality of war and our perception of it. It was almost like watching an NFL game on TV with its tiny figures on the screen compared to being down there on the field in the mud and the blood in the rain getting your socks knocked off. . . . All the potential gains of war without the costs. It could even be mildly entertaining.[38]

The author has seen pilots and sensor operators flying at Holloman Air Force Base and this analogy is true; the technology looks exactly like a video game. Even flying it is similar to a joy stick on a video game. The author is not making the point here that American servicemen and servicewomen should be sacrificed. Rather, the point is that many men, women, and children around the world are faceless, nameless human flesh targets that do not receive a respectful or dignified death. Pilots and sensor operators can see as far as ten feet from their targets. Sometimes, ten feet is just not enough as will be elaborated on throughout the book.

(4) The fourth problem is the numerous mechanical deficiencies in drones that cause them to crash, a lot. Medea Benjamin talks about the mechanical problems and inaccuracy of drones in her book *Drone Warfare*. As a result of mechanical problems, drones can go rogue and take their weapons payload with them often going into highly populous areas. During one instance when a pilot pressed the spacebar on his computer, the drone went into self-destruct mode and a crisis was averted when the drone went into Washington D.C. airspace. In 2009, the Air Force shared that approximately 33 percent of their Predator spy planes had crashed in either Afghanistan or Iraq.[39] Cole, Dobbing, and Hailwood offer a summary of drone mishaps stating that the United States Air Force has had at least seventy-nine drone incidents costing at least one million per incident.[40] Lt. Col. Matt Martin talks about problems with the Predator cockpit. When the human pilot typed an incorrect sequence, it caused the aircraft to engage autopilot when the human pilot was actually trying to turn on the engine cooling fan.[41] Repairs costed over a million dollars in that particular incident. The pilots, sensor operators, and technology have improved but there are still accidents.

There are a few Internet websites that track drone crashes and identify the government proprietor of the drone. One such website is the *Drone Crash Database* that found over eleven USAF drone crashes alone in 2013 and twelve in 2012 from reported sites.[42] Drones are also susceptible to hacking by insurgents thus allowing terrorists to hijack drones for retaliatory sinister purposes; although unlikely, this is possible.[43] Here is a table of all known American crashes and their costs in Table I.1: *American Drone Crashes*. Between 2007 and 2015, the costs of these crashes that we know about has reached approximately 2.6 billion dollars.

(5) The fifth problem with drones is their astronomical costs. As can be seen in Table I.2: *The Costs and Capabilities of American Drones*, the Global Hawk costs 225 million dollars. The MQ-1B Predator costs five million dollars and is one of two lethal drones used on a regular basis. The Predator is in the process of being phased out. The other prevalent lethal drone regularly used is the MQ-9 Reaper, which costs about fourteen million dollars.[44]

Besides their original costs, drones are often damaged by pilots and are expensive to repair. Pilot training is particularly a costly endeavor for the United States Military. Lt. Col. Matt Martin stated that during his twelve

Table I.1 American Drone Crashes[45]

Year	*Crash Victims*	*Costs*	*Total*
2007	A160T-1	4,000,000	94,500,000
	MQ Predator-7	35,000,000	
	MQ-9A Reaper-1	14,000,000	
	MQ5B-Hunter-5	20,000,000	
	RQ-1C Warrior-1	21,500,000	
2008	MQ Predator-12	60,000,000	758,600,000
	MQ-9A Reaper-3	42,000,000	
	MQ5A and B-Hunter-3	9,000,000	
	RQ-1C Warrior-3	645,000,000	
	RQ-1L Gnat-1	2,600,000	
2009	MQ Predator-13	65,000,000	382,100,000
	MQ-9 Reaper-4	56,000,000	
	MQ5A and B-Hunter-4	12,000,000	
	RQ-1C Warrior-1	21,500,000	
	RQ-1L Gnat-1	2,600,000	
	RQ-4A-Global Hawk-1	225,000,000	
2010	MQ Predator-10	50,000,000	111,500,000
	MQ-9 Reaper-2	28,000,000	
	MQ5A and B-Hunter-4	12,000,000	
	RQ-1C Warrior-1	21,500,000	
2011	MQ Predator-14	70,000,000	403,000,000
	MQ-1C Gray Eagle-1	21,500,000	
	MQ-9 Reaper-4	56,000,000	
	MQ5A and B-Hunter-1	3,000,000	
	MQ-170 Sentinel-1	6,000,000	
	RQ-1C Warrior-1	21,500,000	
	RQ-4A-Global Hawk-1	225,000,000	
2012	MQ Predator-10	50,000,000	541,500,000
	MQ-1C Gray Eagle-6	129,000,000	
	MQ-8 Fire Scout-3	54,000,000	
	MQ-9 Reaper-4	56,000,000	
	MQ5A and B-Hunter-2	6,000,000	
	RQ-1C Warrior-1	21,500,000	
	RQ-4A-Global Hawk-1	225,000,000	
2013	KMAX-1	4,000,000	221,400,000
	MQ Predator-8	40,000,000	
	MQ-1C Gray Eagle-3	64,500,000	
	MQ-8 Fire Scout-1	18,000,000	
	MQ-9 Reaper-5	70,000,000	
	MQ5A and B-Hunter-1	3,000,000	
	RQ-1C Warrior-1	21,500,000	
	S100 Camcopter-1	400,000	
2014	MQ Predator-10	50,000,000	92,000,000
	MQ-9 Reaper-3	42,000,000	
2015	MQ Predator-8	40,000,000	69,500,000
	MQ-9 Reaper-1	14,000,000	
	RQ-7B Shadow-1	15,500,000	
Total			**2,674,100,000**

weeks of training at Creech Air Force Base, students scraped up three aircraft tailboards at a replacement cost of ten grand each; crashed one aircraft, destroying it, to the tune of about four million; burned up thousands of gallons of aviation fuel; and shot off eight Hellfire missiles at sixty-eight thousand dollars per missile. I figured Uncle Sam spent a cool quarter million dollars training me to be a Predator pilot.[46]

(6) The sixth problem with targeted killing and drones is that they are a violation of state sovereignty. Henry Waxman, a California House Representative, stated that "While the strike on bin Laden would normally be a violation of state sovereignty, the U.S. government 'is well within its rights' to use force on foreign soil without consent if there is an overriding necessity of self-defense."[47] Drone usage in Pakistan is complex as the CIA faxes drone requests over once per month to the Pakistani Government and gets no response. The United States has decided to call this "tacit response" even though the Pakistani Government has publicly denied that it allows the United States to send these drones.[48] Pakistan later closed its American drone base after some mishaps with unanimous consent of parliament. Most countries that the United States uses targeted killing in have not given their permission for the United States to conduct drone attacks, as will be elaborated, on country-wise, throughout the book. However, the foreign aid that America gives to these nations, in addition to the annihilation of problematic terrorists from time to time, makes it difficult and useless to protest as the United States will do what it wishes anyway.

(7) The seventh problem with drones and targeted killing is the concept of a "legitimate target," which is inherently unclear and central to the debate of targeted killing. In her book, *Legitimate Targets?*, Janina Dill seeks to define "the effectiveness of IHL [International Humanitarian Law] in defining a legitimate target of attack in United States air warfare."[49] Dill finds that IHL cannot reconcile non-pacifist foreign policies with liberal human rights-centered conceptions of appropriateness. . . . Assassinations might be the solution to the dilemma that while war is morally costly so is pacifism. For that to be the case, though, the use of UAV and the legal "permission to assassinate" would need to be regulated and institutionalized, an endeavor for which it is hard to imagine, let alone establish, the requisite international structures. Perhaps nothing short of world government could control and properly administer legalized targeted killings. Similarly only radical institutional change in international relations would mean that the logic of efficiency reserved for defensive wars did not undermine the legal regulation of warfare as such. Even more than the logic of efficiency, targeted killings fail to fulfill their promise of improving the normative implications of the justified use of force.[50]

The problem is that the United States, for obvious reasons, has not defined what a legitimate target is and seems to delineate anyone who seems

Table I.2 The Costs and Capabilities of American Drones[51]

Name	*Maker*	*Cost*	*Purpose*	*Capacity*
Global Hawk	Northrup Grunman	225 Million	High-altitude, long endurance ISR	Speed: 310 knots Wingspan: 40 m Weight: 6,781 kg Payload: 1,360 kg Altitude: 18,288 m Range: 16,000 km
FireScout	Northrup Grunman	18 Million	Helicopter Reconnaissance and Targeting Support	Speed: 110 knots Wingspan: N/A Weight: 940 kg Payload: 272 kg Altitude: 6,1000 m Range: 200 km
Reaper	General Atomics	14 Million	Intelligence Collection, Target Acquisition	Speed: 200 knots Wingspan: 20 m Weight: 2,223 kg Payload: 1,701 kg Altitude: 15,240 m Range: 1,850 km
Predator	General Atomics	5 Million	Armed Reconnaissance, Surveillance, Target Acquisition	Speed: 70 knots Wingspan: 17 m Weight: 512 kg Payload: 204 kg Altitude: 7,600 m Range: 1,250 km
ScanEagle	Boeing and Institu	3 Million	Reconnaissance, Surveillance, Target Acquisition	Speed: 48–70 knots Wingspan: 3 m Weight: 18 kg Payload: camera Altitude: 4,876 m Range: 110 km
Puma	AeroVironment	50,000	Close Range Reconnaissance	Speed: 20–45 knots Wingspan: 3 m Weight: 6 kg Payload: camera Altitude: 152 m Range: 15 km
Switchblade	AeroVironment	50,000	Kamikaze Drone Zone-Fits in Backpack for Close Range	Speed: 55–85 knots Wingspan: NA Weight: 2.5 kg Payload: small explosive Altitude: 500 ft Range: 10 km

Table I.2 The Costs and Capabilities of American Drones—*Continued*

Name	*Maker*	*Cost*	*Purpose*	*Capacity*
Wasp III	AeroVironment	50,000	Reconnaissance and Surveillance (low altitude option)	Speed: 20 knots Wingspan: .72 m Weight: .5 kg Payload: camera Altitude: 150 m Range: 5 km
Raven	AeroVironment	35,000	Reconnaissance and Surveillance (low altitude option)	Speed: 26–52 knots Wingspan: 1.4 m Weight: 2 kg Payload: camera Altitude: 152 m Range: 8–11 km

legitimate at the time as a legitimate target. This is most useful in retrospect when mistakes are made and noncombatants are killed. Again, this problem will be thoroughly discussed throughout the book.

(8) The eighth problem with drones can be defined by the term "blowback." Blowback refers to the unintended consequences of targeted killing. For example, when homes are destroyed and innocents are killed, the resentment, anger, and anti-Americanism that occurs is blowback. Retired General Cartwright stated it well. "We're seeing that blowback" he said at the *Chicago Council on Global Affairs*. "If you're trying to kill your way to a solution, no matter how precise you are, you're going to upset people even if they're not targeted."[52]

In addition to those reasons already provided, Wilner[53] states that targeted killings are not effective because of the international condemnation that occurs when countries use targeted killing as the main counterterrorism policy. Other countries would prefer to see terrorists captured and questioned to garner more information instead of being killed. Alliances and partnerships may be put in jeopardy if there is serious disagreement over the use of this tactic. Militant retaliation in the form of terrorist attacks is another reason why targeted killing is not effective. Wilner also states that possible unknown successors may take over after a leader has been killed. These unknown successors may in fact be more violent than previous leadership.

In summary, the problems with targeted killing outweigh the benefits for many scholars. However, bits of research have produced conclusions here and there that targeted killing is problematic or has its benefits. A comprehensive study of targeted killing has not been performed. This book is the first attempt at a comprehensive study of seven Middle Eastern and African countries, concerning how targeted killing and drone strikes have affected each country.

QUANTITATIVE DATA OPERATIONALIZATION AND TERMINOLOGY

Quantitative data concerning American drone strikes from various sources are used throughout the chapters. Table I.3: *Drone Strike Data*, compiles the data that are used for each chapter. Other variables such as terrorist attacks or suicide bombings are described within each chapter. The author has only included data concerning drone strikes and has been careful to eliminate air strike data as at times this data is combined.

Targeted Killing and Drone Strikes

Drone strikes are a tactic used to complete a targeted killing. As a side note, according to Professor Amos Guiora,[54] University of Utah Law Professor and targeted killing expert, the term "targeted killing" is an Israeli term as the Israelis were the first to use this method. Americans commonly use the term "drone strikes" to talk about killing terrorists through the use of drones. Many other terms are synonymous with the term "targeted killing" such as: signature strikes, long-range hot pursuits, selective targeting, surgical eliminations, decapitation, targeted thwarting, manhunts, active self-defense, pinpoint prevention, and focused disruption in addition to many other terms.[55]

According to Nils Melzer,[56] there are five elements of targeted killing and they are as follows:

1. Targeted killings involve the use of lethal force.
2. Targeted killings are designed to target specifically identified individuals.

Table I.3 Drone Strike Data

Variable	*Source*	*Chapter*
American Drone Strikes	*The Bureau of Investigative Journalism*	Chapter 1: Yemen: The Beginning
American Drone Strikes	*The Bureau of Investigative Journalism*	Chapter 2: Afghanistan: The Land of the Unconquerable
American Drone Strikes	Center for the Study of Targeted Killing	Chapter 3: Pakistan: America's Best Ally and Worst Enemy
American Drone Strikes	Center for the Study of Targeted Killing	Chapter 4: Somalia: Lords of War
American Drone Strikes	*The Bureau of Investigative Journalism*	Chapter 5: Libya: The Elimination of an Eccentric Dictator
American Drone Strikes	AirWars.org	Chapter 6: Iraq and Syria: The Genesis of ISIS

3. Targeted killings are carried out with the deliberate intent to kill the individual in question.
4. Targeted killings are used against individuals not in the physical custody of the state.
5. Targeted killings are carried out by states and their central governments.

Targeted killing is based on the concepts of eradication and deterrence for other terrorists. Targeted killing through the use of drone strikes is a way to deter terrorists from attacking a target. Deterrence "rests on convincing an adversary that the costs of taking a particular action outweigh the potential benefits the action is expected to provide. Deterrence is fundamentally about manipulating another's behavior in a way that suits your needs, interests, or goals."[57]

Sarah Holewinski[58] identifies different types of drone strikes. "Personality strikes" and "signature strikes" are the two kinds of drone strikes that the United States uses. Personality strikes are typically predetermined targets based on some type of kill list. In the early years of the War on Terrorism, cards were used to publicly identify the United States' top military targets in Iraq. This kill list, presented at the supposed "Terror Tuesday" meetings, is highly classified and only available to the president and top military and intelligence officials. This targeted killing process is outlined in the book *The Assassination Complex*, edited by Jeremy Scahill.[59] Once a target is identified, the government has sixty days to assassinate the target.

In contrast, signature strikes are based on observed commonalities among terrorists. The best way to describe this is what Lt. Col. Matt J. Martin talks about in his autobiography, *Predator: The Remote-Control Air War over Iraq and Afghanistan: A Pilot's Story.*[60] Lt. Colonel Martin talked about how the United States' Air Force would use the drones to conduct surveillance in Iraq and Afghanistan. Targets, once identified as doing something suspicious, would be followed until it was confirmed that they were either terrorist sympathizers or members. The targets would then be killed after a certain amount of surveillance that was usually unbeknownst to them. The data used in this book does not differentiate between either signature or personality strikes. Drone strike data is combined for analysis. In addition, the term "targeted killing" is primarily used in this book to talk about an assassination through the use of drone technology. The term "drone strike" will commonly be used to talk about a "targeted killing" unless other otherwise denoted.

UNMANNED AERIAL VEHICLES AND DRONES

This book will mostly use the term "drone" unless an author is quoted within the book using another term. Drones as defined by Jai Galliott[61] consist of the following criteria.

1. Do not have a human operator
2. Are designed to be recoverable (even though they may not be used in a way that renders them such)
3. In a military context, are able to exert their power in order to deliver a lethal or nonlethal payload or otherwise perform a function in support of a military force's objectives.

They include categories of unmanned ground vehicle, unmanned maritime vessels (surface and underwater), UAV and unmanned space vehicles (although the latter category may be considered a subset of the other three).

The term drone strikes will also be used to talk about an attack that is conducted with drones. As stated above, the term "targeted killing" will be used to talk about the drone strikes that are conducted in an effort to destroy a person, place, or thing.

THE ACTORS

Who are the Targeters?

For the most part, there are a few actors involved in the targeting of terrorists. The CIA is usually responsible for the "personality strikes," which are typically determined by intelligence collection. CIA analysts and agents collect data and determine who is of interest. The Special Operations Command's (SOCOM) purpose is to prepare and familiarize Special Forces with a particular area. SOCOM was active in over 100 countries in 2013 and its budget was around 1.8 billion in fiscal year 2012. In 2015, SOCOM received about $10 billion of the $560 billion defense budget, which is about 1.8 percent of DOD's budget.[62] The Defense Department's Joint Special Operations Command (JSOC) is a secret operation and even its budget is classified; SOCOM is the parent organization of JSOC. JSOC plans drone strikes.[63] The Air Force is responsible for carrying out drone attacks. Typically these pilots are located in the Southwest United States bases such as Holloman Air Force Base or Creech Air Force Base and they fly drones located in bases in Afghanistan or Djibouti, etc. to target terrorists in Middle Eastern and African countries.

Who are "the Terrorists"/Targets in These Countries?

Al-Qa'ida

Chapter 1 on Yemen, Chapter 2 on Afghanistan, and Chapter 3 on Pakistan primarily focus on al-Qa'ida. American drone strikes primarily target al-Qa'ida with the Taliban coming in at a close second, depending on which American president's administration one is referring to. Created by a pious Osama bin Laden, son of a Saudi Arabian construction tycoon and a Syrian mother (Alia Ghanem), al-Qa'ida came out of the 1979 revolution in Afghanistan. Osama bin Laden was one of fifty-four children; his father Mohammed bin Awad bin Laden had twenty-two wives, but would divorce and remarry at will to keep the Islamic law of having four wives at one time. Since he was one of the younger sons, it was unlikely that Osama would have been involved in the family construction business. Osama often stated that his father wanted one son to fight for Islam and Osama took it upon himself to fulfill his father's goal.

After a trip to Afghanistan, Osama became appalled at the Soviet invasion and became convinced he should die a martyr repelling the Soviets. Without any military training, Osama created a military base in Jaji, close to a Soviet garrison, which caused the Soviets to frequently bomb the base for hours at a time. On September 10, 1988, al-Qa'ida was created from anti-Soviet elements within Afghanistan. Osama bin Laden moved the headquarters of al-Qa'ida from Afghanistan to Peshwar, Pakistan, after the Saudi Arabian royal family embraced the United States as its protector ally in 1991. By 1992, mujahedeen men in Pakistan were being driven out of the country by the government, fearful of the extremist elements in the mujahedeen population. Osama then went to Sudan to headquarter al-Qa'ida with many of his top advisors. After the Sudanese Government was pressured by the United States to expel bin Laden from Sudan, al-Qa'ida moved back to Afghanistan in 1996. On August 23, 1996, Osama bin Laden declared war on the United States, being embittered by his forced move to Afghanistan.[64]

A string of attacks against the United States followed with the first bombing of the World Trade Center on February 26, 1993. In October 1993, eighteen American servicemen were killed in a humanitarian mission to Mogadishu, Somalia. On August 7, 1998, American embassies in Kenya and Tanzania were bombed. In May 2000, twenty-one hostages were kidnapped at a diving resort by al-Qa'ida members in Malaysia. On October 12, 2000, the *USS Cole* was attacked in Yemen. On September 11, 2001, four hijacked planes flown by nineteen hijackers flew into the United States World Trade Center, the Pentagon, and a field in Pennsylvania. On May 2, 2011, Osama bin Laden was killed by American forces in Abbottabad, Pakistan. His body is buried at sea as no country was willing to take the body or bury it according to Islamic law.[65] However, al-Qa'ida has survived the loss of its founder and continues to grow in numbers.

The Taliban

Chapter 2 on Afghanistan and Chapter 3 on Pakistan focus on the Taliban. Created officially in 1996 by Mullah Mohammed Omar, the Taliban are mostly Pashtuns who live in southern Afghanistan, in the Northwest Frontier Province in Pakistan, or in the Federally Administered Tribal Areas. As the founder of the Islamic Emirate of Afghanistan, Omar has rarely been photographed and refuses to speak to non-Muslims or journalists. Omar is characterized as shy and often allows his commanders to speak for him. Given the title, "Commander of the faithful" like the Prophet Mohammed, Omar makes decisions according to his dreams. Omar once controlled Taliban operations from Kandahar, which he made the capitol of Afghanistan during the Taliban's reign from 1996 to 2001. However, after Hamid Karzai took over in December 2001, Omar has continued to fight the Karzai government from Quetta or Karachi; both are located in Pakistan.[66]

The Taliban typically ignore the border between Afghanistan and Pakistan and cross the border at will. Pakistan and the Pakistani military are either tolerant or supportive of the Taliban. In 2008, sociologist Rosemarie Skaine identified Taliban numbers at approximately 5,000 full-time fighters and 10,000 part-time fighters. The Taliban is governed by a thirty-three person council in addition to a ten-member council of commanders who led military operations.[67]

The Relationship between al-Qa'ida and the Taliban

In 2001, after 9–11, the United States invaded Afghanistan to destroy al-Qa'ida training grounds and, later, the Taliban. The Taliban Government was illegitimate and had allowed al-Qa'ida and other terrorist elements to train in Afghanistan. The relationship must be briefly discussed between the Taliban and al-Qa'ida from a current perspective. The United States has denounced the Taliban as a terrorist group and often targets them with drones.

Anne Sternerson[68] talks about the nature of the relationship between al-Qa'ida and the Taliban after 2001 and the invasion of Afghanistan. She states that both groups consist of like-minded individuals who are each beholden to centralized leadership within their own sphere of influence; however, neither group is homogenous. Individual groups can act on their own when necessary.

The relationship between al-Qa'ida varies tremendously based on the country and region. At times al-Qa'ida has married Pashtun women in Afghanistan and Pakistan to solidify relationships with tribes and secure loyalty. On other occasions, Taliban members have spoken of their dislike for al-Qa'ida members and the danger that often accompanies members of al-Qa'ida. At times, al-Qa'ida will supply troops and weapons for Taliban operations. Al-Qa'ida currently resides in the eastern and southeastern areas of Afghanistan where they have better relationships with the Taliban.

ISIS/ISIL

Chapter 6 (Iraq and Syria: The Genesis of ISIS) looks at the effects of targeted killing on ISIS/ISIL in Syria and Iraq. In its original conception, this book was supposed to have a chapter on Syria and a chapter on Iraq. However, when looking at the data on targeted killing in both Iraq and Syria, it occurred to the author that most of the targeted killing transpired for the purposes of destroying some component of ISIS. ISI (Iraqi Islamic State) came out of AQI or al-Qa'ida in Iraq and this group later announced its affiliation with al-Qa'ida in Syria in May 2013. After the two branches united, the entire Iraqi and Syrian group became known as ISIS or ISIL, depending on one's favored term and perspective. ISIS refers to the Islamic State in Iraq and Syria. ISIL refers to the Islamic State of Iraq and the Levant. The Levant is defined as the lands around Syria or "lands of the rising sun," including Jordan, Israel, Palestine, and Syria.

Al-Shabaab

Chapter 4 on Somalia focuses on al-Shabaab who is al-Qa'ida's affiliate in Somalia. Al-Shabaab was created in 2007 as an offspring of the defeated Union of Islamic Courts in Somalia. Based on Wahhabism and Sunni in origin, al-Shabaab is dedicated to making an Islamic State in Somalia and to restoring the Caliphate. Somewhere around 2010, perhaps as late as 2012, al-Shabaab declared their allegiance to al-Qa'ida. Slowly but surely, al-Shabaab has managed to take over large swaths of territory and has committed major attacks not only in Somalia but also in Kenya.[69] Current estimates put al-Shabaab's numbers at somewhere between 6,000 and 12,000 members.[70]

Effective Counterterrorism

Since this book looks at the effects of drone strikes and targeted killing on the countries that experience them, effective counterterrorism in this book refers to several specifications. They are as follows:

1. Effective counterterrorism requires a decrease in terrorism in the countries where drone strikes occur whether categorized as terrorist attacks or suicide terrorist attacks. An increase in drone strikes should then have a decrease in terrorist attacks in each country.
2. Effective counterterrorism does not create blowback in which radicalization occurs, anti-Americanism is created, and terrorist group recruitment and empathy increases.
3. Effective counterterrorism has a collateral damage rate of less than 10 percent. This means from a human being perspective that for every

terrorist that is killed, there must be one civilian killed or for every ten terrorists that are killed, approximately one civilian is killed.

For targeted killing to be an effective foreign policy tool, the previous three criteria must be met throughout the book. The summary of the data from each chapter will be discussed in the conclusion and the effectiveness of targeted killing in the Middle East and Africa will be evaluated. This includes the examination of whether targeted killing has been effective in the decrease of terrorism in the United States. The purpose of the War on Terrorism is to protect the United States and to decrease terrorism and terrorist groups throughout the world. George W. Bush stated, "Our war on terror begins with al-Qa'ida, but it does not end there. It will not end until every terrorist group of global reach has been found, stopped and defeated."[71] The conclusion will also examine comprehensive costs of drone warfare, which thus far have been proclaimed as cheaper than traditional warfare.

AN OVERVIEW OF THE BOOK

The book begins with Chapter 1 (Yemen: The Beginning), which looks at the first country to experience American drone strikes both in the Middle East and in the world. The chapter focuses on al-Qa'ida in Yemen and early American drone strikes procedures. In Chapter 2 (Afghanistan: The Land of the Unconquerable), the true spirit of Afghanis who are the most heavily droned country in the world is discussed. Chapter 3 (Pakistan: America's Best Ally and Worst Enemy) looks how Pakistan has been both an ally and an enemy in the War on Terror, often playing a double-sided game with al-Qa'ida and the Taliban. In Chapter 4 (Somalia: Lords of War), American foreign policy is discussed concerning al-Shabaab and the civil war in Somalia that has been raging in the failed state for decades. Chapter 5 (Libya: The Elimination of an Eccentric Dictator) is unique in that it looks at the small campaign the United States conducted in cooperation with NATO to rid the world of Qaddafi. The last Chapter 6 (Iraq and Syria: The Genesis of ISIS) looks at the current war with ISIS in both Iraq and Syria and whether drone warfare and targeted killing have been successful in defeating ISIS. The concluding chapter provides a summary of all the data concerning targeted killing in the Middle East and Africa assesses targeted killing as a counterterrorism policy in that region of the world. The concluding chapter also discusses whether targeted killing has made the United States and Americans safer. Lastly, the conclusion quantifies known fiscal costs of drone warfare and then examines the costs of traditional warfare in Iraq and Afghanistan in comparison. The purpose of this is to provide a preliminary assessment as to whether drone warfare is really more cost effective than traditional warfare.

NOTES

1. Peter Bergen and Jennifer Rowland, "Decade of the Drone," in *Drone Wars*, eds. Peter L. Bergen and Daniel Rothenberg (Cambridge: Cambridge University Press, 2015), 15.

2. Peter Bergen and Jennifer Rowland, "Decade of the Drone," in *Drone Wars*, eds. Peter L. Bergen and Daniel Rothenberg (Cambridge: Cambridge University Press, 2015), 16.

3. Jon Boone, "US limits Pakistan drone strikes amid political battle over military moves," *The Guardian,* February 5, 2014, accessed February 5, 2014, http://www.theguardian.com/world/2014/feb/05/us-limits-pakistan-drone-strikes.

4. In a hearing before the Committee on Armed Services, Dr. David Kilcullen from the Center for a New American Security stated that the Pakistani people "have not been brought into it [drone warfare in Pakistan] and that they should be consulted about drones and targeted killing in Pakistan." House Hearing 111 Congress, "Effective Counterinsurgency: The Future of the U.S.-Pakistan Military Partnership," Hearing before the full committee of the Committee on Armed Services April 23, 2009, accessed October 15, 2014, http://www.gpo.gov/fdsys/pkg/CHRG-111hhrg52666/content-detail.html.

5. George W. Bush, "President Bush Addresses the Nation," *The Washington Post*, September 20, 2001, accessed August 26, 2015, http://www.washingtonpost.com/wp-srv/nation/specials/attacked/transcripts/bushaddress_092001.html.

6. Avery Plaw, *Targeting Terrorists A License to Kill?* (Cornwall, England: Ashgate, 2008).

7. James DeShaw Rae, *Analyzing the Drone Debates: Targeted Killing, Remote Warfare, and Military Technology* (New York: Palgrave Macmillan, 2014).

8. Although the military prefers the term UAV or unmanned aerial vehicle, this book will use the term drone. UAV is preferable for the military as it sounds better (according to interviews).

9. Benjamin Wittes and Gabriella Blum, *The Future of Violence* (New York: Basic Books, 2015), 5.

10. David Axe, *War Bots, U.S. Military Robots Are Transforming War in Iraq, Afghanistan, and the Future* (Ann Arbor, MI: Nimble Books LLC, 2008).

11. Daniel Byman, "Why Drones Work," *Foreign Affairs* 92 (32) (2013).

12. Alex S. Wilner, *Deterring Rational Fanatics* (Philadelphia: University of Pennsylvania Press, 2015), 95–108.

13. Patrick B. Johnston and Anoop K. Sarbahi, "The Impact of US Drone Strikes on Terrorism in Pakistan and Afghanistan," *Princeton University*, July 14, 2013, accessed September 30, 2014, https://webspace.princeton.edu/users/esocweb/ESOC%20website%20publications/JS_DronesTerrorism.pdf

14. Medea Benjamin, *Drone Warfare Killing by Remote Control* (New York: Verso, 2013), 28.

15. Audrey Kurth Cronin, "Why Drones Fail, When Tactics Drive Strategy," *Foreign Affairs* 92 (44) (2013).

16. Medea Benjamin, *Drone Warfare Killing by Remote Control* (New York: Verso, 2013), 106.

17. Approximately 52% of Pakistanis believed in the accuracy of American drone strikes in 2009. See Farhat Taj, "Drone Attacks-A Survey," *International News*, March 5, 2009, accessed January 28, 2015, http://www.thenews.com.pk/TodaysPrintDetail.aspx?ID=165781&Cat=9&dt=3/4/2009.

18. See Lt. Col. Matt J. Martin with Charles W. Sasser, *Predator: The Remote-Control Air War over Iraq and Afghanistan: A Pilot's Story* (Minneapolis: Zenith Press, 2010) for numerous examples of human collateral damage.

19. Brian Glyn Williams, *Predators, The CIA's Drone War on al Qaeda* (Dulles, Virginia: Potomac Books, 2013), 166–168.

20. Lucy Draper, "Report Accuses Israel of Human Rights Abuses in Gaza During Summer War," *Newsweek*, January 22, 2015, accessed August 4, 2016, http://www.newsweek.com/report-accuses-israel-human-rights-abuses-during-summer-war-301275.

21. This definition is part of an article published in 2008 by Christine Sixta Rinehart in *The Journal of Women, Politics, and Policy*. The title of the article is "The Illusive Third Wave: Are Female Terrorists the New "New Women" In Developing Societies?"

22. State-sponsored terrorism is terrorism that is paid for by a government. Payment may come in many forms such as weaponry, money, or political protection. Saddam Hussein sponsored terrorism in Israel through Palestinian fighters. Although different countries may sponsor terrorist groups in this paper manner they are violent for their own purposes regardless of where their money comes from. This definition applies only to the terrorist organizations I have chosen to study within this book. State-sponsored terrorism is an extremely complicated issue that deserves separate scholarship apart from nongovernmental terrorism.

23. Christine Sixta Rinehart, *Volatile Social Movements and the Origins of Terrorism: The Radicalization of Change.* (Lanham, MD: Lexington Books, 2012), Introduction.

24. The White House, "Executive Order," Whitehouse.gov, July 1, 2016, accessed August 4, 2016, https://www.whitehouse.gov/the-press-office/2016/07/01/executive-order-united-states-policy-pre-and-post-strike-measures.

25. Paul McLeary with Adam Rawnsley, "White House Drone Report," *The Situation Report, The National Security Daily Brief* from Foreign Policy, July 1, 2016, accessed August 4, 2016, http://link.foreignpolicy.com/view/54da7bb010defb923d8b49ec47b0k.1gd6/2052ebeb.

26. Jeremy Scahill, *Dirty Wars: The World is a Battlefield* (New York: Nation Books, 2013), 78.

27. Jeremy Scahill, *Dirty Wars: The World is a Battlefield* (New York: Nation Books, 2013), 78.

28. See also Lloyd C. Gardener, *Killing Machine, The American Presidency in the Age of Drone Warfare* (New York: The New Press, 2013), 209–210.

29. Joseba Zulaika, "Drones and Fantasy in US Counterterrorism," *Journal for Cultural Research*, 18:2, 2014: 171–187.

30. United States Department of Justice, "White Paper," *Department of Justice*, February 4, 2013, accessed March 29, 2016, http://msnbcmedia.msn.com/i/msnbc/sections/news/020413_DOJ_White_Paper.pdf.

31. Medea Benjamin, *Drone Warfare Killing by Remote Control* (New York: Verso, 2013), 123.

32. The fifth, six, and eighth Amendments guarantee that no American citizen shall be deprived of life, liberty, or property without due process, a right to a speedy and public trial, and no cruel and unusual punishments. So far 8 American citizens have been killed by drones: Anwar al-Awlaki, Abdulrahman al-Awlaki, Samir Khan, Jude Kenan Mohammad, Kamal Derwish, Ahmed Farouq, Adam Gadahn, and Warren Weinstein. There is disagreement as to whether most of these men were propagandists or immediate threats. Weinstein was an accident.

James DeShaw Rae, *Analyzing the Drone Debates: Targeted Killing, Remote Warfare, and Military Technology* (New York: Palgrave Macmillan, 2014), 105.

See also Lloyd C. Gardener, *Killing Machine, The American Presidency in the Age of Drone Warfare* (New York: The New Press, 2013).

33. Jeremy Scahill, *Dirty Wars: The World is a Battlefield* (New York: Nation Books, 2013), 1–2.

34. US Department of Justice, "Attorney General Eric Holder Speaks at Northwestern University School of Law," Department of Justice, Chicago, IL, March 5, 2012.

35. Peter Baker, "In Terror Shift, Obama Took a Long Path," *The New York Times*, May 27, 2013, Accessed August 21, 2015, http://www.nytimes.com/2013/05/28/us/politics/in-terror-shift-obama-took-a-long-path.html?_r=0.

36. Jeffrey S. Brand and Amos Guiora, "Establishment of a Drone Court: A Necessary Restraint on Executive Power," in *The Legitimacy of Drones: UCAVs for Cross-Border Counterterrorism,* ed. S Barela (Burlington, VT: Ashgate, 2015).

37. Lt. Col. Matt J. Martin with Charles W. Sasser, *Predator: The Remote-Control Air War over Iraq and Afghanistan: A Pilot's Story* (Minneapolis: Zenith Press, 2010), 19.

I have also been to Holloman Air Force Base on September 10, 2015, where I witnessed Air Force Pilots and Sensor Operators flying drones in tractor trailers. This is essentially for the purpose of easily moving the trailers and technology in them into helicopters and naval ships. The technology could easily be moved across the world, being encompassed within a trailer.

38. Lt. Col. Matt J. Martin with Charles W. Sasser, *Predator: The Remote-Control Air War over Iraq and Afghanistan: A Pilot's Story* (Minneapolis: Zenith Press, 2010), 46–47.

39. Medea Benjamin, *Drone Warfare Killing by Remote Control* (New York: Verso, 2013), 22.

40. Chris Cole, Mary Dobbing, and Amy Hailwood, "Convenient Killing Armed Drones and the 'Playstation' Mentality," *The Fellowship of Reconciliation*, September 2010, accessed September 30, 2014, http://dronewarsuk.files.wordpress.com/2010/10/conv-killing-final.pdf.

41. Lt. Col. Matt J. Martin with Charles W. Sasser, *Predator The Remote-Control Air War over Iraq and Afghanistan: A Pilot's Story* (Minneapolis: Zenith Press, 2010), 197–198.

42. *Drone Crash Database*, "Drone Wars UK," March 31, 2014, accessed October 15, 2014, http://dronewars.net/drone-crash-database/.

43. Chris Cole, Mary Dobbing, and Amy Hailwood, "Convenient Killing Armed Drones and the 'Playstation' Mentality," *The Fellowship of Reconciliation*, September 2010, accessed September 30, 2014, http://dronewarsuk.files.wordpress.com/2010/10/conv-killing-final.pdf.

See the following for drones crashed by Hezbollah.

Drone Crash Database, "Drone Wars UK, March 31, 2014," accessed October 15, 2014, http://dronewars.net/drone-crash-database/.

44. *Stanford International Human Rights & Conflict Resolution Clinic*, "Drones: An Overview," 2012, accessed October 12, 2014, http://www.livingunderdrones.org/background-and-context/.

45. *Drone Crash Database*, "Drone Wars UK," March 31, 2014, accessed October 15, 2014, http://dronewars.net/drone-crash-database/.

46. Lt. Col. Matt J. Martin with Charles W. Sasser, *Predator: The Remote-Control Air War over Iraq and Afghanistan: A Pilot's Story* (Minneapolis: Zenith Press, 2010), 26.

47. Jonathan Masters, "Targeted Killings," *Council on Foreign Relations*, May 23, 2013, accessed July 14, 2014, http://www.cfr.org/counterterrorism/targeted-killings/p9627.

48. Adam Entous, Siobhan Gorman, and Evan Perez, "U.S. Unease Over Drone Strikes," *The Wall Street Journal*, September 26, 2012, accessed October 12, 2014, http://online.wsj.com/news/articles/SB100008723963904441004045776415208580114 52.

49. Janina Dill, *Legitimate Targets? Social Construction, International Law and US Bombing* (Cambridge: Cambridge University Press, 2015), 7.

50. Janina Dill, *Legitimate Targets? Social Construction, International Law and US Bombing* (Cambridge: Cambridge University Press, 2015), 309.

51. James DeShaw Rae, *Analyzing the Drone Debates: Targeted Killing, Remote Warfare, and Military Technology* (New York: Palgrave MacMillan, 2014), 11.

United States Air Force, accessed September 11, 2014, http://www.af.mil.

P.W. Singer, *Wired for War: The Robotics Revolution and Conflict in the Twenty-first Century* (New York: Penguin Press, 2009).

52. Mark Mazetti and Scott Shane, "As New Drone Policy Is Weighed, Few Practical Effects Are Seen," *The New York Times*, March 21, 2013, accessed August 2015, http://www.nytimes.com/2013/03/22/us/influential-ex-aide-to-obama-voices-concern-on-drone-strikes.html.

53. Alex S. Wilner, *Deterring Rational Fanatics* (Philadelphia: University of Pennsylvania Press, 2015), 108–116.

54. Amos Guiora, "Executive Power and Terrorism," panel presented at Wofford College, Spartanburg, SC, March 28, 2016.

55. Alex S. Wilner, *Deterring Rational Fanatics* (Philadelphia: University of Pennsylvania Press, 2015), 76.

56. Nils Melzer, *Targeted Killing in International Law* (Oxford: Oxford University Press, 2008), 3–6.

57. Alex S. Wilner, *Deterring Rational Fanatics* (Philadelphia: University of Pennsylvania Press, 2015), 16.

58. Sarah Holewinski, "Just Trust Us, The Need to Know More about the Civilian Impact of US Drone Strikes," in *Drone Wars*, eds. Peter L. Bergen and Daniel Rothenberg (Cambridge: Cambridge University Press, 2015), 45–47.

59. Jeremy Scahill, ed., *The Assassination Complex* (New York: Simon & Schuster, 2016).

60. Lt. Col. Matt J. Martin with Charles W. Sasser, *Predator: The Remote-Control Air War over Iraq and Afghanistan: A Pilot's Story* (Minneapolis: Zenith Press, 2010).

61. Jai Galliot, *Military Robots, Mapping the Moral Landscape* (Burlington, Vermont: Ashgate Publishing Company, 2015), 5.

62. Marcus Weisgerber, "Peeling the Onion Back on the Pentagon's Special Operations Budget," Defense One, January 27, 2015, accessed May 26, 2016, http://www.defenseone.com/management/2015/01/peeling-onion-back-pentagons-special-operations-budget/103905/.

63. Laurence Pope, *The Demilitarization of American Diplomacy* (New York: Palgrave Macmillan, 2014), Chapter 4.

64. Peter L. Bergen, *The Longest War, The Enduring Conflict between America and al-Qa'ida* (New York: Free Press, 2011).

65. CNN Wire Staff, "Timeline: Osama bin Laden, Over the Years," *CNN*, May 2, 2011, accessed December 30, 2015, http://www.cnn.com/2011/WORLD/asiapcf/05/02/bin.laden.timeline/.

66. Peter Bergen and Katherine Tiedemann, Editors, *Talibanistan, Negotiating the Borders between Terror, Politics, and Religion* (Oxford: Oxford University Press, 2013).

67. Rosemarie Skaine, *Women of Afghanistan in the Post-Taliban Era, How Lives Have Changed and Where They Stand Today* (Jefferson, NC: McFarland & Company, Inc., Publishers, 2008), 34.

68. Anne Stenersen, "The Relationship between al-Qa'ida and the Taliban," in *Talibanistan*, eds. Peter Bergen and Katherine Tiedemann (New York: Oxford University Press, 2013), 69–93.

69. Daniel Byman, *Al Qaeda, The Islamic State, and The Global Jihadist Movement, What Everyone Needs to Know* (Oxford: Oxford University Press, 2015), 146.

70. Alexander Meleagrou-Hitchens and Abdihakim Ainte, "The Return of al-Shabaab," *The Daily Beast*, March 15, 2016, accessed on May 25, 2016, http://www.thedailybeast.com/articles/2016/03/15/the-return-of-al-shabaab.html.

71. George W. Bush, "President Bush Addresses the Nation," *The Washington Post*, September 20, 2001, accessed August 26, 2015, http://www.washingtonpost.com/wp-srv/nation/specials/attacked/transcripts/bushaddress_092001.htm

Chapter 1

Yemen

The Beginning

The first American drone strike ever occurred on November 3, 2002, a few hours outside of Sanaa, Yemen. Two Hellfire missiles targeted the speeding car containing Abu Ali al-Harithi,[1] a suspected al-Qa'ida planner in the terrorist attack on the *USS Cole* and six of his companions, one being an American citizen. Although the first missile was unsuccessful and skidded into the sand, the second missile found its mark. Five people were killed in the attack. Unbeknownst to the US Air Force pilot flying the drone in the United States at the time, Abdul Rauf Nassib survived. Abdul Rauf Nassib was later arrested in Yemen in 2004 for helping ten al-Qa'ida members (who assisted in the *USS Cole* attack) escape from Aden Prison in 2003.[2] This first drone strike was the only drone strike that President George W. Bush ordered in Yemen, although the order actually came from CIA Director Gorge Tenet and Lt. General Michael DeLong.[3] President Obama, according to his own confessions, has ordered every drone strike since November 3, 2002, in Yemen and has increased the rate of drone strikes in Yemen astronomically. In a new book, *Double Down,* affirmed by several sources, President Obama stated, "Turns out I'm really good at killing people. Didn't know that was gonna be a strong suit of mine" in reference to his drone program.[4] President Obama has also given numerous speeches on Iraq, Syria, in addition to other countries where he states that he personally ordered strikes.

The current Yemeni political situation, and the relationship concerning American targeted killing, requires elaboration and discussion. Yemen consists of approximately five boiling cauldrons, all waiting to create explosions or implosions, depending on the issue. Some might argue that indeed the cauldrons have recently exploded in early 2015 as Yemen is currently seething with more violence and economic stagnation than almost any state in the world. The first of these cauldrons are the economic challenges. Even though oil is a major export of Yemen, the oil resources are dwindling, possibly completely depleted by 2020. Water, already a scarce resource, faces a 7 percent population growth rate. *Qat*, an addictive cultural staple, chewed by 75 percent of Yemini men and some women, requires more water to grow than other crops. Food staples have been forgone in an effort to grow *Qat,* so Yemen has little food production. Corruption, inflation, and an economic downturn add to the previously mentioned economic issues. The second cauldron consists of around two-thirds of the population being under twenty-four years old and unable to find employment. Over 50 percent of the population is illiterate, where women are approximately 70 percent illiterate. The third cauldron is domestic security issues. Al-Qa'ida and other Islamic terrorist groups create violence, death, and destruction almost daily. A civil war has been occurring in Saada since 2004, where the government is fighting the Zaidi Shi'i revivalists who want to make Yemen a theocracy. There is also a secessionist movement in South Yemen. The fourth cauldron is the inability of Yemen to secure its border so the smuggling of drugs, weapons, and other vast arrays of illegal items move across Yemen regularly. The last cauldron is Yemen's proximity to Somalia where piracy is a daily reality and Somalian violence spills over into Yemen.[5]

Thrown in with these five boiling cauldrons is a recent political situation that is strange than fiction. Ali Abdullah Saleh Al-Snhani Al-Humairi ruled Yemen in some capacity since 1978, ruling North Yemen and then the entire Yemen since 1990 after the country's unification. Saleh's nickname was simply "the boss" as his ability to maneuver political obstacles was unprecedented in a difficult country where tribal ties were paramount. Saleh is known for collaborating with anyone he needs at the moment and is somewhat of a deal-maker.[6] To better make the point, Saleh once stated that governing Yemen was like, "dancing on the heads of snakes," which eventually became an excellent book title about Saleh's life.[7] In February 2012, President Saleh's rule came to an abrupt end when he was thrown out of office by democratic protestors. His Vice President and hand picked successor Abd Rabbuh Mansur Hadi quickly filled Saleh's shoes until his own oust by Mohammed Ali al-Houthi in February 2015; since then Yemen has been plagued by a violent civil war complete with the intervention of numerous international actors all vying for their own interests.

Most countries refuse to recognize al-Houthi as the president and still maintain that President Hadi is the rightful president, even though he no longer controls Sana'a, the capital of Yemen and has returned to and fled from Yemen several times. Saudi Arabia led an international attempt, complete with numerous allies to try to restore President Hadi's authority. However this quest was discontinued in April 2015 as the civilian death toll climbed dramatically due to the Saudi Air Force bombings. Currently, there appears to be a stalemate between the two sides and President Hadi is now living in

Saudi Arabia. Mercenaries from numerous countries and even the American Company Blackwater, now known as Academie, have been involved in the civil war.

But the part that is stranger than fiction is that former President Saleh made a deal with al-Qa'ida shortly before he was thrown out of office to give the Southern province of Yemen to al-Qa'ida. Saleh even withdrew the military from Abyan, within the Southern province, giving al-Qa'ida easy access without impediment. Saleh even allowed the Western-trained anti-terror unit, led by his nephew to quietly dissolve. The Houthi uprising has occurred for many reasons, one of the major reasons being that al-Qa'ida has taken over their home territory and slowly taken over the Southern part of Yemen. Saleh has recently switched sides again, fighting with the Iranian-backed Houthis to restore the country, stating that he is the only one who can keep Yemen free from the clutches of al-Qa'ida. Although the United States had suspicions that Saleh, who was once an ally, had made deals with al-Qa'ida, he was relatively trustworthy and none of these claims were substantiated. Eventually, the United States figured out that President Saleh was never what he seemed to be and withdrew their support of him.

Although American targeted killing in Yemen has been occurring since 2002, both the current and former presidents of Yemen have shown semi-cooperative attitudes toward the American policy, depending on the time frame. President Saleh allowed the United States to use Yemen air bases in return for monetary aid and military arms. Later this relationship was described as nothing more than a "gentleman's agreement," not an official alliance.[8] Others described the relationship as "passive," stating that the United States did what it wanted in Yemen and passed a few crumbs over to Saleh and Hadi for their compliance.[9] After President Hadi came to power, the CIA requested an expansion of the drone campaign in Yemen, the request coming from General David Petraeus.[10] Current ousted President Hadi has also strongly supported the American drone policy, stating in September 2012, "They pinpoint the target and have zero margin of error, if you know what target you're aiming at." The United States, "helped with their drones because the Yemeni Air Force cannot carry out missions at night. The electronic brain's precision is unmatched by the human brain."[11] President Hadi stated that he personally approved every drone strike that the United States sent. Until his ousting, President Hadi was praised by the United States as one of their most supportive counterterrorism allies in the War on Terror.[12] It is evident that both the CIA and JSOC are actively targeting drones in Yemen without any real sanctioned consent at the moment as no one is officially in charge of the country.[13]

The Houthis, currently in control of Yemen however, are not as cooperative with the American drone policy. In early September 2015, the Houthi rebels attacked a military air base that is currently being used by coalition forces and had previously been used by Americans to launch drone strikes in Yemen.[14] Mohammad al-Bukhaiti, a member in the Houthi political office in Sana'a, stated that the Houthis "condemn US operations in Yemen and any violation of Yemeni sovereignty." In response, the current government is coercing the National Security Bureau to withhold intelligence from the Americans. "We are well aware that US drone strikes do not target Al-Qa'ida members exclusively, but other people for political reasons as well," al-Bukhaiti added. "This is helping spread support for Al-Qa'ida. Yemenis sympathize with the group when they see their brothers getting killed by Americans."[15]

AL QA'IDA IN YEMEN

Al-Qa'ida has had a presence in Yemen since approximately 1997 or May 1998,[16] although it was not until 2002 that the Bush Administration committed its first drone targeted killing in Yemen against Qaed Salim Sinan al-Harethi, a wanted al-Qa'ida member. After following him from populated areas, al-Harethi's car was destroyed and six men were killed including one American citizen as previously mentioned. All that was left was a black stain in the sand.[17] While the Yemeni Government secretly allowed this strike to occur, it was betrayed when US Government officials publicized the act as a way to get support for the War on Terrorism. What followed was a US withdrawal of support from the Yemeni Government and the United States also withdrew twenty million in aid and 140 million in World Bank aid.

Since 2002, al-Qa'ida has continued to grow in Yemen, partly from other al-Qa'ida members immigrating to the country for safety. Former American Ambassador Barbara Bodine stated, "By about 2004, we and the Yemenis had the conventional al-Qa'ida problem reasonably under control. We then took our eye of the ball [with] Iraq."[18] It is commonly accepted that al-Qa'ida revived itself in 2006 as approximately twenty-three prisoners escaped from the Yemeni Political Security Organization, which was probably an inside job. In reality, six of the prisoners were killed, eleven returned to prison and only six prisoners remained at large in the population.[19] It is more likely that the major cause of al-Qa'ida revitalization is the merger that occurred in January 2009, when al-Qa'ida cells in Saudi Arabia and cells in Yemen united, thus creating al-Qa'ida in the Arabian Peninsula (AQAP). As a result, the Yemeni Government rounded up 170 suspects and made them pledge not to engage in terrorist activities: the men were then put under the supervision of their tribal elders.[20] Founding members included Nasir al-Wuhayshi (served with bin Laden in Afghanistan), Said Ali Al-Shiri and Mohamed Atiq Awayd al-Harbi (both former Guantanamo Bay detainees') in addition to others from

Algeria, Bosnia, and Chechnya. In 2011, al-Qa'ida began moving down to the Southern part of Yemen from the mountains and taking over cities.

Although still loyal to al-Qa'ida, AQAP wants to create its own Islamic emirate in Yemen. AQAP has a three-part strategy that it uses to build its envisioned emirate.[21] The first part is the purposeful exploitation of tribal conflict. AQAP will take members from the territory of one tribe and insert those members into another tribe's territory, guaranteeing that any conflict will lead to tribal retaliation if the outsider is harmed. Second, al-Qa'ida continues to increase its territorial control in some cases by actually governing a region, and al-Qa'ida loots and attacks noncompromising regions. The third part involves indoctrinating the local population to increase support and membership such as supporting the South's secession from Yemen in 2009. In addition, AQAP constantly undermines the legitimate government in Yemen.

Oil has become one of al-Qa'ida's favorite targets in Yemen in addition to foreign embassies, foreigners, and security officials. One of al-Qa'ida's major goals is to control the Bab al-Mandab Strait in Yemen to control the territory between Yemen and Somalia. If al-Qa'ida controls this shipping waterway, they are able to control goods coming in and out of Somalia and Yemen, thus tremendously increasing their revenue in Yemen.

Al-Qa'ida has also harnessed the power of Saudi citizens in AQAP with Yemeni tribal ties to solidify Yemen and Saudi Arabian unity. This is exemplified in a letter from Abu Basir al-Wuhayshi, the former leader of al-Qa'ida in Yemen who was killed in an American drone strike in June 2015, entitled "To Our People in the South" on May 13, 2009:

> I hereby say to the people of the south, the south of the Arabian Peninsula, avoiding all colonial division of our countries, from the Yamani side to the Arabian Sea, to our proud tribes, who God has favored with Islam and made them occupy the world, I say to them that I am proud of being one of them . . . I tell them that our country is in the middle of the world, we have the Qiblah, Mecca and Medina. We have the richest seas, the most important straits and the greatest reserves of oil. Do you accept that the crusaders and their agents control you? Do you accept that they steal your money and your resources? Would you allow them to kill your sons and daughters? Oh people of the Arabian Peninsula, oh people of the Arabian Peninsula, you are the strategic depth of your fellow jihadists.[22]

The population of al-Qa'ida in Yemen is ethnically diverse. Yemenis are 56 percent of AQAP, while Saudis comprise approximately 37 percent of the organization; foreigners make up the other 7 percent of AQAP. AQAP is based mostly on tribal ties and incorporates tribes in the north and south of Yemen.[23] Although al-Qa'ida is the dominant terrorist group within Yemen Ansar al-Shari'ah, another terrorist group in Yemen is used as the mouthpiece for AQAP to create solidarity and to unite the local populace.[24]

The Yemeni Government believes that the United States exaggerates the strength of AQAP believing that there are "significant al-Qa'ida populations in Yemen and Somalia."[25] However, Academic Christina Hellmich disagreed in 2011 stating that "To date, as both recent (attempted) attacks and statement in the name of AQAP indicate, the group's activity appears to be at a new peak: it pursues an ambitious agenda that both local and global, aspiring to attack the enemy-the USA and its allies, the Yemini government and other Arab leaders-in Yemen and abroad."[26] Yemini journalist Abdalilah Haydar Shay'a, who regularly interviews AQAP states that the realistic number of AQAP is more likely around 3000.[27]

THE EFFECTS OF AMERICAN TARGETED KILLING IN YEMEN

In Figure 1.1: *American Drone Strikes and Maximum Deaths, Yemen 2002–2015,* American drone strikes in Yemen and resulting deaths are graphed. This data is taken from *The Bureau of Investigative Journalism.* Maximum deaths are used as opposed to minimum deaths as maximum deaths are typically reported by the local people who are familiar with and perhaps were at the scene of the strike. In addition, maximum deaths include people who were hit and later died from their injuries. In Yemen, approximately 7.3 people are killed for every drone strike, which would include the militants in this number as we are not sure who is a militant and who is not. We are also unsure as to who is targeted. The total number of people killed in Yemen from 2002 until the end of 2015 are 1269, most of them recently since 2011. The total number of drone strikes is 174. It is highly unlikely that over seven people are targeted per drone strike. Most likely the number is closer to two or three. The inordinate number of innocent people killed per drone strike in Yemen is

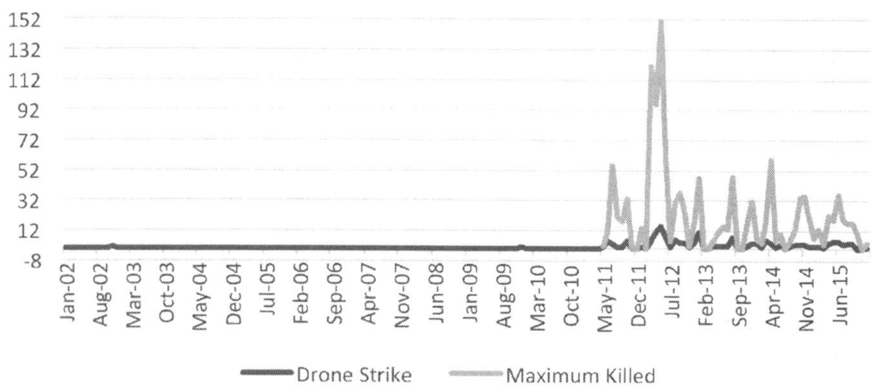

Figure 1.1 American Drone Strikes and Maximum Deaths, Yemen 2002–2015.[28]

unacceptable and leads to incredible insurmountable challenges for the United States such as an increase in al-Qa'ida members.

THE SUCCESSES IN YEMEN

A few high-profile members of al-Qa'ida have been assassinated by drones in Yemen. The first man to be targeted in Yemen on November 3, 2002, was Abu Ali al-Harithi,[29] a suspected al-Qa'ida planner in the terrorist attack on the *USS Cole* and six of his companions, one being an American citizen. Shortly after this drone strike occurred, the United States invaded Iraq in March 2003. In response to the drone strike and the US invasion of Iraq, Yemini citizens by the "tens of thousands" began protesting in the streets and marched on the American embassy. Three people were killed and dozens were injured.[30]

The second major target was Anwar al-Awlaki, an American and Yemini citizen. Although he was charged with being a recruiter in al-Qa'ida, the story is not that simple. Al-Awlaki was an imam and knew three of the 9–11 hijackers, Nidal Malik Hasan (Fort Hood), and Umar Farouk Abdulmutallab (Attempted Christmas Day bombing). In his early years, al-Awlaki was the moderate "go to" guy for media comments concerning violent jihadists. He condemned the actions of the jihadists. As time went on, his connections with the 9–11 hijackers from his local mosque made him an FBI target. It is not known whether al-Awlaki was always a sleeper cell, who was radicalized by the United States' Government, or whether he was radicalized over time. His secrets and his life were destroyed when he was assassinated by a drone in Yemen on September 30, 2011. A few weeks later, his sixteen-year-old son Abdulrahman, also an American, was killed by a drone strike in Yemen after the young man went searching for his father.[31] Samir Khan, the Editor of online al-Qa'ida magazine, was killed in the same drone strike as Anwar al-Awlaki.[32]

Sheikh Abu Zubeir 'Adil al'Abab, the sharia official, described as AQAP's fourth-most important leader, was killed in an American drone strike on October 4, 2012.[33] Sa'id Ali Jabir Al Khathim Al Shihri, a deputy leader of AQAP and a former Guantanamo detainee was killed by a drone strike in early 2013. Al-Shihri was responsible for funding and for AQAP's offensive strategy.[34] Ibrahim al-Rubaish was killed in a drone strike on April 12, 2015. Al-Rubaish was an al-Qa'ida mufti and was a former Guantanamo detainee, having spent five years at the Detention Center. He escaped from Saudi Arabian authorities in 2006 after being released from Guantanamo and the State Department placed a five million dollar bounty on his head.[35] Nasir al-Wayshi was killed on June 12, 2015, in Yemen; he was one of the founder members of AQAP. Al-Wayshi was handpicked to take over for Ayman al-Zawahiri and was immediately replaced with Qasm al-Rimi.[36]

By the author's count, in summary, less than ten high-profile al-Qa'ida members have been killed since 2002, when the targeted killings began in Yemen. These are considered "high-profile terrorists" because media sources published their deaths. According to *The Bureau of Investigative Journalism*, approximately 1269 people have been killed in pursuit of the elimination of these approximate seven high-profile al-Qa'ida terrorists. Most assuredly, several low-level militants have been killed also. Low-level militants or terrorists are difficult to ascertain as no evidence exists as to who the people actually were and the United States does not publish their strikes or kills. This is where it becomes murky as to who a terrorist is.

Yemenis are concerned as to how people are being targeted. Some of those who have been assassinated such as Anwar al-Awlaki were killed for making speeches, there was never any direct link to al-Qa'ida except his statements. A court of law never tried these people and no charges were brought against these citizens before they were targeted.

THE MAJOR OOPSIES

In Yemen, there are have been several huge "mistakes," where civilians were targeted by American drones and not one terrorist was found among the dead or in the location. The following are just a few of the larger mistakes that have been made in Yemen. The first of these occurred on December 17, 2009, and was the worst with over forty-three dead civilians. Among those killed were twenty-two children and five pregnant women.[37] Although this was not a drone strike, but a missile launched from a US Navy ship, it still affects the American relations with the Yemeni population. It is not likely that Yemenis are concerned about whether their loved ones are killed with a drone versus a missile strike.

On January 23, 2012, a drone strike outside of Sana'a killed a university student Saleem Hussein Jamal and his cousin Ali Nasser Jamal. The two men had stopped when five men had offered to pay them for a ride. A lethal mistake was made by Jamal and his cousin as the five men were al-Qa'ida members. One witness to the crash site stated, "We found eyes, but there were no faces left."[38]

On August 29, 2012, a Yemini cleric named Salem Ahmed bin Ali Jaber stood up to make a speech in his village condemning al-Qa'ida. Two days later, Jaber was confronted by al-Qa'ida members concerning his speech. Along with his cousin Waleed Abdullah (a police officer brought for protection) Mr. Jaber confronted the three al-Qa'ida members. All five people were killed in a drone strike in addition to a camel that was nearby. Mr. Jaber was an opponent of al-Qa'ida and was killed for protecting his village from al-Qa'ida influence.[39] His family later received 100,000 U.S. dollars in a plastic bag from a Yemini Government official but no apology was ever made by President Obama.[40] The family later went to the United States and sued the American Government, only wanting President Obama to apologize and admit to the deaths of Waleed and Jaber.[41]

On September 20, 2012, outside Radda, a Toyota truck, full of fourteen people, was targeted by an American drone strike. Eleven people were killed including a woman, a seven-year-old girl, and a twelve-year-old boy. None of the victims were identified as al-Qa'ida, although the Yemeni Government stated that the drones were targeting al-Qa'ida member Abdel-rauf al-Dahab, in an attempted cover up for the American Government. Furious tribesmen took the bodies to the presidential palace to confront the government.

Another mistake occurred on December 12, 2013, when four Hellfire missiles attacked a wedding convoy of eleven vehicles outside Rad'a, Yemen, killing twelve men and injuring fifteen. The bride was also injured in the attack, receiving a facial wound and her trousseau was blown to pieces. Her stepson was also one of the men who were assassinated. In response, angry citizens blocked the main road to Rad'a for over twenty-four hours, displaying the bodies of the dead until government officials gave 20 Kalashnikov rifles and thirty-two Yemeni Riyals (blood money worth 150,000 US dollars), a culturally relevant apology to the families. Neither Yemen nor the United States took public responsibility for the action. The United States advocated that they were targeting a major militant Shawqi Ali Ahmad al-Badani, who escaped the attack but no had ever heard of him.

Human Rights Watch investigated and received several statements from those who were present during the same attack. One elderly man present at the attack stated, "We were in a wedding, but all of a sudden it became a funeral. . . . We have nothing, not even tractors or other machinery. We work with our hands. Why did the United States do this to us?" Another man shouted at the protests that later occurred, "The US is committing bloodshed! Who will stand against it?" Abdullah Muhammad Al-Tisi was among several residents present who demanded an international investigation be conducted, stating, "The US government killed innocent people. This was an immense mistake, rejected by the God and His Prophet. They turned many kids into orphans, many wives into widows. Many were killed and many others injured. . . . Where is the concern for human rights? Our skies should be clear of these drones. We can't lead our lives. We can't sleep. After this attack, men, women, and children became more afraid; they are afraid of another strike. It has become a nightmare."[42]

Other villagers threatened retaliation by the al-Qaifah tribal confederation if they did not see prosecutions and/or receive additional compensation.

Abdu Rabu Abdullah al-Tisi, who lost four relatives in the strike, told *Human Rights Watch*: "If the government makes good on its word . . . then we're okay. But our tribe is very big, and it will not forget the blood of our sons; it will not let this blood flow in vain."[43]

The wedding attack spurred the condemnation of parliament. The Yemeni parliament had a vote late in December 2013 to ban all American drone strikes in Yemen. Although parliament's votes are non-binding to the president, the action was taken to stop American drone strikes in Yemen. The Yemen state news agency SABA stated, "Members of parliament voted to stop what drones are doing in Yemeni airspace, stressing the importance of preserving innocent civilian lives against any attack and maintaining Yemeni sovereignty."[44]

AMERICAN DRONE STRIKES INCREASE RADICALIZATION IN YEMEN

There are several articles that give quotes from citizens who have lost loved ones killed by American drone strikes. These Yemeni citizens routinely state that they are likely to join or support al-Qa'ida as a result of the drone strikes. An article from *The New York Times,* by author Ibrahim Monthana, finds that drone strikes are increasing radicalization in Yemen. Monthana, co-founder of the political party, called the Wantan Party in Yemen stated that "Drone strikes are causing more and more Yemenis to hate America and join radical militants; they are not driven by ideology but rather by a sense of revenge and despair."[45] In the article, Monthana talks about a Yemini lawyer who states, "DEAR OBAMA, when a U.S. drone missile kills a child in Yemen, the father will go to war with you, guaranteed. Nothing to do with Al Qaeda." Monthana adds that al Qa'ida in 2009 only had a few hundred members and controlled no land. Today, al-Qa'ida along with Ansar al-Sharia has around 3,000 members and controls a considerable amount of Yemeni land.

As referred to previously in *The Major Oopsies* section on September 2012, where 11 civilians were killed without any proof of a terrorist presence, the families of those victims were radicalized by the drone strikes. "The two survivors and relatives of six victims, interviewed separately and speaking to a Western journalist about the incident for the first time, expressed willingness to support or even fight alongside AQAP, as the al Qaeda group is known."[46] Mohammed said, "Our entire village is angry at the government and the Americans. If the Americans are responsible, I would have no choice but to sympathize with al-Qa'ida because al-Qa'ida is fighting America." The driver in the attack Nasser Mabkhoot Mohammed al-Sabooly stated, "If we are ignored and neglected, I would try to take my revenge. I would even hijack an army pickup, drive it back to my village and hold the soldiers in it hostages. I would fight along al-Qa'ida's side against whoever was behind the attack."[47]

Al-Qa'ida provides incentives for people to join. One man stated, "Al Qaeda always gives money to the family," said Hussein Ahmed Othman al Arwali, a tribal sheikh. "Al Qaeda's leaders may be killed by drones, but the group still has its money, and people are still joining. For young men who are poor, the incentives are very strong: they offer you marriage, or money, and the ideological part works for some people."[48]

In an article from *The Washington Post*, May 29, 2012, Sudarsan Raghavan uses interviews of over 20 people to make the case that the drone strikes are increasing al-Qa'ida recruitment and sympathy in Yemen. One businessman Salim al-Barakani stated, "These attacks are making people say, 'We believe now that al-Qa'ida is on the right side, adding that his two brothers—one a teacher, the other a cellphone repairman—were killed in a U.S. strike in March'." Mohammed al-Ahmadi, legal coordinator for Karama, a local human rights group stated "Every time the American attacks increase, they increase the rage of the Yemeni people, especially in al-Qa'ida-controlled areas. The drones are killing al-Qa'ida leaders, but they are also turning them into heroes." Another man Aidaroos stated, "The Americans are targeting the sons of the Awlaki. I would fight even the devil to exact revenge for my nephew." Abdelaziz Muhammed Hamza, head of the Revolutionary Council in Abyan province, a group that is fighting AQAP stated that, "All the residents of the area have joined al-Qa'ida."[49] Human rights worker Abdul Rahman Berman, the executive director of the National Organization for Defending Rights and Freedoms or Hood, puts it best when he stated, "The drone war is failing. If the Americans kill 10, al-Qa'ida will recruit 100."[50]

Christopher Swift published an article in *Foreign Affairs* based on the opinions of 40 older, conservative Yemen tribal leaders, Islamist politicians, Salafist clerics, and other sources living in 14 of Yemen's 21 provinces. In this article, only five of the people interviewed believed that drone strikes were helping al-Qa'ida more than they were hurting it. Instead the men cited the economic situation as reasons for young men to join al-Qa'ida. Al-Qa'ida has jobs and gives young men cars, cash, and rifles. Ordinary Yemenis view drones as an affront to their national sovereignty. "Drones remind us that we don't have the ability to solve our problems by ourselves," one member of the Yemeni Socialist Party said. "If these were Yemeni drones, rather than American drones, there would be no issue at all." Yemini politicians stated that, "No one resents a drone strike if the target was a terrorist," a member of the Muslim Brotherhood stated. "What we resent is the fact that outsiders are involved."[51]

Preeminent Yemen scholar Gregory Johnsen believes that American drone strikes have caused around an increase in AQAP recruitment. Johnsen states in an interview,

> In the West, the debate over U.S. policy in Yemen has become focused on drone strikes, but in Yemen, the focus is on the civilian casualties that are a result of some of those strikes. When the Obama Administration started carrying out attacks in Yemen, there were about 200–300 individuals affiliated with AQAP. Today, it's at least 1,000—in fact, the U.S. State Department estimates that it's at least a few thousand. I don't think all of this is attributable to the use of drones, or to the civilian casualties they've resulted in, but I think a large portion of it is, and because of this, one of the things that I think the U.S. has to do is reconsider its strike policy.[52]

Numerous of articles have appeared concerning al-Qa'ida recruitment and sympathy in Yemen as a result of American drone strikes. Al-Qa'ida is proficient at setting up training camps in the countryside and using local towns for supplies and recruitment of young boys. In southern Yemen, in the town of al-Mahfad, residents asked al-Qa'ida to leave after suffering three years of abuse. The only reason that the camp was finally asked to leave was because *Rueters* published articles indicating where the camp was located. Al-Qa'ida was involved in mounting an insurgency against the Yemini military and would recruit from a local population with high unemployment.[53]

Another unintended consequence is the migration of Yemenis who are afraid of living in al-Qa'ida areas, because it is like a prison and there are constant fears of drone strikes. Many of the migrants have moved south to Aden where a large population remains living in squalor and garbage as there is nowhere else to go. However, al-Qa'ida continues to take over towns and advance its way across Yemen.[54]

QUANTITATIVE ANALYSIS

In Figure 1.2: *American Drone Strikes and Terrorist Attacks in Yemen, 2011–2015*, American drone strikes and al-Qa'ida terrorist attacks are shown in a line graph. This data is charted by month for each variable. Using bivariate regression, it was found that there was no statistically significant relationship between American drone strikes and terrorist attacks in Yemen. This can be further explained by stating that American drone strikes do not cause an increase or decrease in terrorist attacks nor do terrorist attacks lead to an increase or decrease in drone strikes. Contrary to the original purpose in the War on Terror, American drone strikes are not decreasing or increasing terrorism in Yemen. In fact, there is no relationship. The drone strike data were

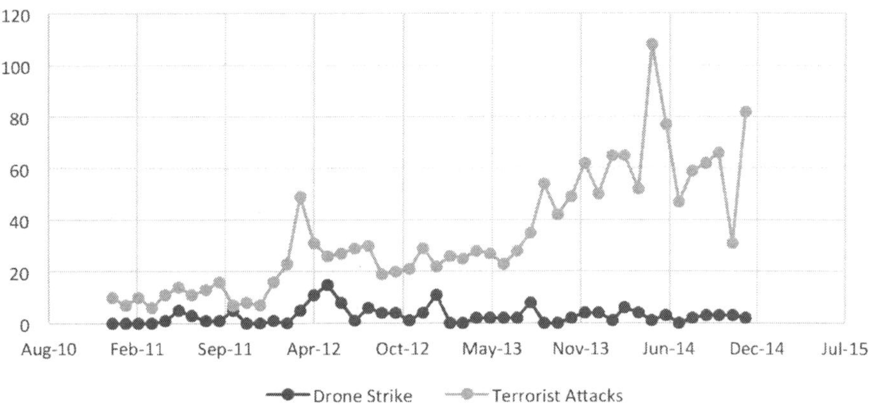

Figure 1.2 American Drone Strikes and Terrorist Attacks in Yemen, 2011–2015. *Source*: *The Bureau of Investigative Journalism*, "US Strikes in Yemen, 2002-present" 2015, https://docs.google.com/spreadsheets/d/1lb1hEYJ_oml8lSe33izwS2a2lbiygs0hTp2Al_Kz5KQ/edit and Chicago Project on Security and Terrorism, The University of Chicago, 2015, http://cpostdata.uchicago.edu/search_results_new.php.

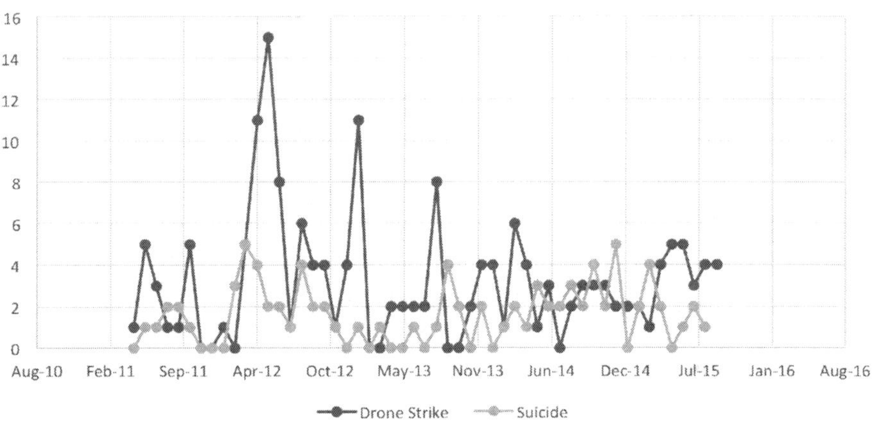

Figure 1.3 American Drone Strikes and Suicide Bombings in Yemen, 2011–2015. *Source*: *The Bureau of Investigative Journalism*, "US Strikes in Yemen, 2002-present" 2015, https://docs.google.com/spreadsheets/d/1lb1hEYJ_oml8lSe33izwS2a2lbiygs0hT-p2Al_Kz5KQ/edit and Chicago Project on Security and Terrorism, The University of Chicago, 2015, http://cpostdata.uchicago.edu/search_results_new.php.

taken from *The Bureau of Investigative Journalism*[55] and the al-Qa'ida terrorist attack data were taken from *The Chicago Project for Security and Terrorism*.[56]

Although the data appears to be somewhat correlated in the graph Figure 1.3: *American Drone Strikes and al-Qa'ida Suicide Bombings in Yemen 2011–2015*, quantitative statistical analysis did not prove to be statistically significant using simple bivariate regression. The drone strike data were taken from *The Bureau of Investigative Journalism*[57] and the suicide bombing data were taken from *The Chicago Project for Security and Terrorism.*[58] This means that American drone strikes do not have any measurable effect on suicide bombings in Yemen nor do suicide bombings lead to more or less drone strikes in Yemen. From another perspective, this can be interpreted as suicide terrorist attacks do not increase or decrease drone strikes in Yemen, which is somewhat counterintuitive to the idea that drone strikes are supposed to decrease terrorism.

Although it is evident above that drone strikes have killed a few major al-Qa'ida leaders since 2002, replacements have quickly been made and al-Qa'ida does not seem to be running out of manpower anytime soon. It is also evident that al-Qa'ida does not stress the need for suicide terrorist attacks in Yemen like they do in Pakistan. This is most likely due to the difference in al-Qa'ida branches and leadership.

CONCLUSION

In conclusion, the targeted killings in Yemen appear disorganized, counterproductive, and with high amounts of collateral damage. Here are three more examples of problematic drone strikes in Yemen. In his book, *Dirty Wars*, Jeremy Scahill writes about Abdulelah Haider Shaye, a Yemini journalist who was imprisoned by Yemeni and Americans forces for exposing the truth in Yemen about drone warfare. Shaye was accused of being an al-Qa'ida propagandist, although those that know him deny that he could ever be involved with al-Qa'ida. He was held and tortured for several days and was sentenced to five years in a Yemeni prison. When the Yemeni Government wanted to free Shaye from prison, President Obama voiced his concern and it was not until 2013 that Shay was released.[59]

The second example consists of Adnan al Qahdi who was killed in a drone strike in November 2012. Al Qadhi was a public supporter of al-Qa'ida but was also a mediator for the Yemeni Government concerning their negotiations with al-Qa'ida. Mr. Qadhi was even earning a government salary at the time of his death. Although he could have been arrested as he never made any attempt to hide and was living in the open, Mr. Qadhi was killed in a drone strike.[60] To add to this example, an eight-year-old boy was recruited by Yemeni Republican Guard to plant tiny electric chips on Adnan Qahdi, the boy's surrogate father. The question here becomes, did the United States assist in the recruitment of an eight-year-old boy to help assassinate his surrogate father?[61]

Since the War on Terrorism began, several jihadists have been arrested in Yemen for no crime other than jihadist beliefs as these men were considered dangerous to society. The Committee for Dialogue was established in Yemen to ascertain whether these young men should be rehabilitated into society. The process consisted of an introduction to a moderate Yemeni cleric who would access the detainee's interpretation of religious scriptures. If the detainee could convince the cleric of the legitimacy of their personal jihad, the detainee would be let go. However, this did not happen and many young men were set straight by the true Islam. This program has been moderately successful. Several hundred of these young men have been released. Besides assassination, rehabilitation is the only other option for jihadists although freely expressing their opinions are a jail sentence.[62]

In addition, if drone strikes are used to assassinate "immediate threats," why are more of these "terrorists" not arrested such as Adnan Qahdi, who was completely accessible to authorities at any time? Yemen has an elite counterterrorism unit that was trained by the United States and was formed in 2003 to confront Yemen's terrorism threats. This elite unit was created by John Brennan and yet the unit is never used.[63] One officer in the unit stated, "For sure, we could be going after some of these guys. That's what we're trained to do, and the Americans trained us. It doesn't make sense."

Drone strikes are not the major problem. It is high number of drone strikes, the death of innocent civilians, and the drone strikes with incorrect intelligence that are causing problems in Yemen. It is likely that Yemenis would be more supportive of the American drone strikes if high-ranking al-Qa'ida members were killed and very few, if any civilians, were killed. The technology of the drone is highly sophisticated, and accurate and the pilots are well trained. There is no excuse for mistakes with such highly accurate technology. The truth of the matter is the United States is cutting off its nose to spite its face in countries like Yemen. Although it is impossible to quantify the number of al-Qa'ida members at any given time, experts agree that al-Qa'ida has grown and that drone strikes have not deterred nor depleted the terrorist organization in any aspect.

NOTES

1. Abdul Rauf Nassib was the bodyguard of Mohammad Hamdi al-Ahdal who helped to orchestrate the attack on the French oil tanker *MV Limburg* in October 2002.

2. Gregory D. Johnsen, *The Last Refuge* (New York: W.W. Norton & Company, 2013), 122–123; *BBC News*, "Yemen Captures al-Qa'ida Members," *BBC News*, March 4, 2004, accessed June 1, 2015, http://news.bbc.co.uk/2/hi/middle_east/3531657.stm.

3. Transcript 6, "Top Secret America," *PBS Frontline*, 2010, accessed January 5, 2016, http://www.pbs.org/wgbh/frontline/film/topsecretamerica.

4. Mollie Reilly, "Obama Told Aides He's 'Really Good At Killing People,' New Book 'Double Down' Claims," *The Huffington Post,* November 4, 2013, accessed August 17, 2016, http://www.huffingtonpost.com/2013/11/03/obama-drones-double-down_n_4208815.html.

5. Christopher Boucek, "Yemen: Avoiding a Downward Spiral," in *Yemen on the Brink*, eds. Christopher Boucek and Marina Ottaway (Washington D.C.: Carnegie Endowment for International Peace, 2010), 1–27.

6. Jeremy Scahill, *Dirty Wars: The World is a Battlefield* (New York: Nation Books, 2013), 61.

7. Sanaa, "Newsmaker: Yemen's Saleh dance 'on heads of snakes' nears end," *Reuters*, November 23, 2011, accessed January 11, 2015, http://www.reuters.com/article/us-yemen-saleh-idUSTRE7AM1UR20111123.

8. Letta Tayler, "A Wedding that Became a Funeral," *Human Rights Watch*, February 19, 2014, accessed January 5, 2016, https://www.hrw.org/report/2014/02/19/wedding-became-funeral/us-drone-attack-marriage-procession-yemen.

9. Gregory D. Johnsen, "Yemen's Passive Role in the War on Terrorism," in *The Battle for Yemen*, ed. Ramsey Mardini (Washington D.C.: The Jamestown Foundation, 2010), 87–90.

10. Greg Miller, "CIA seeks new authority to expand Yemen drone campaign," *The Washington Post*, April 18, 2012, accessed December 30, 2015, https://www.washingtonpost.com/world/national-security/cia-seeks-new-authority-to-expand-yemen-drone-campaign/2012/04/18/gIQAsaumRT_story.html.

11. Scott Shane, "Yemen's Leader Praises U.S. Drone Strikes," *The New York Times*, September 29, 2012, accessed December 20, 2015, http://www.nytimes.com/2012/09/29/world/middleeast/yemens-leader-president-hadi-praises-us-drone-strikes.html.

12. Richard Spencer, "How Yemen's US-backed ex-dictator is tearing his country apart," *The Telegraph*, March 28, 2015, accessed December 30, 2015, http://www.telegraph.co.uk/news/worldnews/middleeast/yemen/11502027/How-Yemens-US-backed-ex-dictator-is-tearing-his-country-apart.html.

13. Noah Shachtman and Spencer Ackerman, "Let's Admit It: The U.S. is at War in Yemen Too," *Wired*, June 14, 2012, accessed December 30, 2015, http://www.wired.com/2012/06/yemen-war/.

14. Kareem Fahim, "Houthi Rebels Kill 45 U.A.E. Soldiers in Yemen Fighting," *The New York Times*, September 4, 2015, accessed December 30, 2015, http://www.nytimes.com/2015/09/05/world/middleeast/27-soldiers-in-saudi-led-force-reported-killed-in-yemen.html?_r=0.

15. Ali Ibrahim al-Moshki, "Three US Drone Strikes in Eight Days," *Yemen Times*, February 4, 2015, accessed December 30, 2015, http://www.yementimes.com/en/1857/news/4860/Three-US-drone-strikes-in-eight-days.htm.

16. Amb Abdalla and Arezou Hassanzadeh, "Al-Qa'ida and Terrorism in the Arab East: Rise Decline, and the Effects of Doctrine Revisions and the Arab Revolutions," in *Al-Qa'ida After Ten Years of War A Global Perspective of Successes, Failures, and Prospects* edited by Norman Cigar and Stephanie E. Kramer, 100–127. (Quantico, Virginia: Marine Corps University Press, 2011), 103.

17. Aki Peritz and Eric Rosenbach, *Find, Fix, Finish, Inside the Counterterrorism Campaigns That Killed Bin Laden and Devastated Al-Qa'ida* (New York: Public Affairs, 2012), Chapter 10.

18. Azmat Khan, "Understanding Yemen's Al Qaeda Threat," *Frontline*, May 29, 2012, accessed January 14, 2016, http://www.pbs.org/wgbh/frontline/article/understanding-yemens-al-Qa'ida-threat/.

19. W. Andrew Terrill, *The Conflict in Yemen and U.S. National Security* (Carlisle, PA: Strategic Studies Institute, 2011).

20. W. Andrew Terrill, *The Conflict in Yemen and U.S. National Security* (Carlisle, PA: Strategic Studies Institute, 2011).

21. Christopher Swift, "The Boundaries of War? Assessing the Impact of Drone Strikes in Yemen," in *Drone Wars*, eds. Peter L. Bergen and Daniel Rothenberg (Cambridge: Cambridge University Press, 2015), 76–78.

22. Fares Shuwail al-Zahrani, "To Our People in the South," May 2004, accessed December 29, 2015, www.ukht-benladn.net/mhadrat_show-446.html.

23. Murad Batal al-Shishani, "An Assessment of the Anatomy of al-Qa'ida in Yemen: Ideological and Social Factors," *Terrorism Monitor*, VIII (9) (March 5, 2010), accessed December 29, 2015, http://www.jamestown.org/programs/tm/single/?tx_ttnews%5Btt_news%5D=36111&tx_ttnews%5BbackPid%5D=457&no_cache=1#.VoLWDi_Slhc.

24. Christopher Swift, "The Boundaries of War? Assessing the Impact of Drone Strikes in Yemen," in *Drone Wars*, eds. Peter L. Bergen and Daniel Rothenberg (Cambridge: Cambridge University Press, 2015), 71–88.

25. Committee on Foreign Relations, "Al Qaeda in Yemen and Somalia: A Ticking Time Bomb," A Report to the Committee on Foreign Relations United States Senate, January 21, 2010, accessed December 29, 2015, http://www.foreign.senate.gov/imo/media/doc/Yemen.pdf.

26. Christina Hellmich, *Al-Qa'ida from Global Network to Local Franchise* (Black Point, Nova Scotia: Fernwood Publishing ltd, 2011), 145.

27. As cited in Alistair Harris, "Exploiting Grievances: Al-Qa'ida in the Arabian Peninsula," in *Yemen on the Brink*, ed. Christopher Boucek and Marina Ottaway (Washington D.C.: Carnegie Endowment for International Peace, 2010), 44.

28. *The Bureau of Investigative Journalism*, "US Strikes in Yemen, 2002-present" 2015, https://docs.google.com/spreadsheets/d/1lb1hEYJ_omI8lSe33izwS2a2lbiygs0hTp2Al_Kz5KQ/edit.

29. Abdul Rauf Nassib was the bodyguard of Mohammad Hamdi al-Ahdal who helped to orchestrate the attack on the French oil tanker *MV Limburg* in October 2002.

30. Robert F. Worth, "10 are Killed in Bombings at Embassy in Yemen," *The New York Times,* September 17, 2008, accessed January 12, 2016, http://www.nytimes.com/2008/09/18/world/middleeast/18yemen.html?_r=0.

31. Jeremy Scahill, *Dirty Wars: The World is a Battlefield* (New York: Nation Books, 2013), Chapter 2.

32. Jeremy Scahill, *Dirty Wars: The World is a Battlefield* (New York: Nation Books, 2013), Chapter 40.

33. Murad Batal al-Shishani, "Adel al-Abbab: Al-Qa'ida in the Arabian Peninsula's Religious Ideologue," in *The Battle for Yemen*, ed. Ramsey Mardini (Washington D.C.: The Jamestown Foundation, 2010), 151–155.

34. Murad Batal al-Shishani, “From Yemen to Detroit: The Expanding Influence of AQAP’s Sa’id al Shihri,” in *The Battle for Yemen*, ed. Ramsey Mardini (Washington D.C.: The Jamestown Foundation, 2010), 143–147.

35. Murad Batal al-Shishani, “Ibrahim al-Rubaish: New Religious Ideologue of al-Qa’ida in Saudi Arabia Calls for Revival of Assassination Tactic,” in *The Battle for Yemen*, ed. Ramsey Mardini (Washington D.C.: The Jamestown Foundation, 2010), 137–139.

36. Rafid Fadhil Ali, “AQAP’s Man in the South: Nasir al-Wuhayshi,” in *The Battle for Yemen*, ed. Ramsey Mardini (Washington D.C.: The Jamestown Foundation, 2010), 156–159.

Jethro Mullen, “Al Qaeda’s second in command killed in Yemen strike; successor named,” *CNN,* June 16, 2015, accessed January 12, 2016, http://www.cnn.com/2015/06/16/middleeast/yemen-aqap-leader-killed/.

37. Chris Woods, “Drone strikes in Yemen: The civilian massacre the US neither confirms nor denies,” *The Bureau of Investigative Journalism*, March 29, 2012, accessed January 12, 2015, https://www.thebureauinvestigates.com/2012/03/29/the-civilian-massacre-the-us-will-neither-confirm-nor-deny/.

38. Robert F. Worth, Mark Mazzetti, and Scott Shane, “Drone Strikes’ Risks to Get Rare Moment in the Public Eye,” *The New York Times*, February 5, 2013, accessed January 12, 2015, http://www.nytimes.com/2013/02/06/world/middleeast/with-brennan-pick-a-light-on-drone-strikes-hazards.html?_r=0.

39. Robert F. Worth, Mark Mazzetti, and Scott Shane, “Drone Strikes’ Risks to Get Rare Moment in the Public Eye,” *The New York Times*, February 5, 2013, accessed January 12, 2015, http://www.nytimes.com/2013/02/06/world/middleeast/with-brennan-pick-a-light-on-drone-strikes-hazards.html?_r=0.

40. Faisal bin ali al Jaber, “What is the value of an innocent life, Mr. Obama: $100K in a plastic bag?” *CNN Opinion*, June 12, 2015, accessed January 14, 2016, http://www.cnn.com/2015/06/12/opinions/drone-victims-relatives-sue-us-opinion/.

41. Scott Shane, Families of Drone Strike Victims in Yemen File Suit in Washington, *The New York Times,* June 8, 2015, accessed January 14, 2016, http://www.nytimes.com/2015/06/09/world/middleeast/families-of-drone-strike-victims-in-yemen-file-suit-in-washington.html?_r=0.

42. Letta Tayler, “A Wedding that Became a Funeral,” *Human Rights Watch*, February 19, 2014, accessed January 5, 2016, https://www.hrw.org/report/2014/02/19/wedding-became-funeral/us-drone-attack-marriage-procession-yemen.

43. Letta Tayler, “A Wedding that Became a Funeral,” *Human Rights Watch*, February 19, 2014, accessed January 5, 2016, https://www.hrw.org/report/2014/02/19/wedding-became-funeral/us-drone-attack-marriage-procession-yemen.

44. Mohammed Ghobari, “Yemeni Parliament in Non-binding Vote against Drone Strikes,” *Reuters*, December 15, 2013, accessed January 5, 2016, http://www.reuters.com/article/us-yemen-drones-idUSBRE9BE0EN20131215.

45. Ibrahim Monthana, “How Drones Help al-Qa’ida,” *The New York Times*, June 13, 2012, accessed September 23, 2014, http://www.nytimes.com/2012/06/14/opinion/how-drones-help-al-Qa’ida.html?_r=0.

46. Sudarsan Raghavan, “When U.S. drones kill civilians, Yemen’s government tries to conceal it,” *The Washington Post*, December 24, 2012, accessed January

13, 2016, https://www.washingtonpost.com/world/middle_east/when-us-drones-kill-civilians-yemens-government-tries-to-conceal-it/2012/12/24/bd4d7ac2-486d-11e2-8af9-9b50cb4605a7_story.html.

47. Sudarsan Raghavan, "When U.S. drones kill civilians, Yemen's government tries to conceal it," *The Washington Post*, December 24, 2012, accessed January 13, 2016, https://www.washingtonpost.com/world/middle_east/when-us-drones-kill-civilians-yemens-government-tries-to-conceal-it/2012/12/24/bd4d7ac2-486d-11e2-8af9-9b50cb4605a7_story.html.

48. Robert F. Worth, Mark Mazzetti, and Scott Shane, "Drone Strikes' Risks to Get Rare Moment in the Public Eye," *The New York Times*, February 5, 2013, accessed January 12, 2015, http://www.nytimes.com/2013/02/06/world/middleeast/with-brennan-pick-a-light-on-drone-strikes-hazards.html?_r=0.

49. Sudarsan Raghavan, "In Yemen, U.S. airstrikes breed anger, and sympathy for al-Qa'ida," *The Washington Post*, May 29, 2012, accessed October 9, 2014, http://www.washingtonpost.com/world/middle_east/in-yemen-us-airstrikes-breed-anger-and-sympathy-for-al-Qa'ida/2012/05/29/gJQAUmKI0U_story_2.html.

50. Sudarsan Raghavan, "When U.S. drones kill civilians, Yemen's government tries to conceal it," *The Washington Post*, December 24, 2012, accessed January 13, 2016, https://www.washingtonpost.com/world/middle_east/when-us-drones-kill-civilians-yemens-government-tries-to-conceal-it/2012/12/24/bd4d7ac2-486d-11e2-8af9-9b50cb4605a7_story.html.

51. Christopher Swift, "The Drone Blowback Fallacy Strikes in Yemen Aren't Pushing People to Al Qaeda," *Foreign Affairs*, July 1, 2012, accessed October 12, 2014, http://www.foreignaffairs.com/articles/137760/christopher-swift/the-drone-blowback-fallacy.

52. Open Canada Staff, "Gregory Johnsen on Yemen, the U.S., and Drones," *Open Canada.org*, December 14, 2012, accessed January 5, 2016, https://www.opencanada.org/features/gregory-johnsen-on-yemen-the-u-s-and-drones.

53. *Rueters*, "Abandoned Al Qaeda notebook found in Yemen features blueprint for creating Islamic state," July 8, 2014, accessed October 10, 2014, http://www.nydailynews.com/news/national/abandoned-al-Qa'ida-notebook-features-blueprint-creatiwww.goong-islamic-state-article-1.185853477.

54. Ghaith Abdul-Ahad, "Al-Qa'ida's Wretched Utopia and the Battle for Hearts and Minds," *World News*, April 30, 2012, January 14, 2016, http://www.theguardian.com/world/2012/apr/30/alqaida-yemen-jihadis-sharia-law.

55. *The Bureau of Investigative Journalism*, "US Strikes in Yemen, 2002-present," 2015, https://docs.google.com/spreadsheets/d/1lb1hEYJ_omI8lSe33izwS2a2lbiygs0hTp2Al_Kz5KQ/edit.

56. *Chicago Project on Security and Terrorism*, The University of Chicago, 2015, http://cpostdata.uchicago.edu/search_results_new.php.

57. *The Bureau of Investigative Journalism*, "US Strikes in Yemen, 2002-present" 2015, https://docs.google.com/spreadsheets/d/1lb1hEYJ_omI8lSe33izwS2a2lbiygs0hTp2Al_Kz5KQ/edit.

58. *Chicago Project on Security and Terrorism*, The University of Chicago, 2015, http://cpostdata.uchicago.edu/search_results_new.php.

59. Jeremy Scahill, *Dirty Wars: The World is a Battlefield* (New York: Nation Books, 2013), Chapter 41.

60. Robert F. Worth, Mark Mazzetti, and Scott Shane, "Drone Strikes' Risks to Get Rare Moment in the Public Eye," *The New York Times*, February 5, 2013, accessed January 12, 2015, http://www.nytimes.com/2013/02/06/world/middleeast/with-brennan-pick-a-light-on-drone-strikes-hazards.html?_r=0.

61. Gregory D. Johnsen, "Did an 8-Year-Old Spy for America?" *The Atlan*tic, September 2013, accessed January 13, 2016, http://www.theatlantic.com/magazine/archive/2013/09/did-an-8-year-old-spy-for-america/309429/.

62. Michael Taarnby, "Yemen's Committee for Dialogue: Can Jihadists Return to Society?" in *The Battle for Yemen*, ed. Ramsey Mardini (Washington D.C.: The Jamestown Foundation, 2010), 78–82.

63. Lauren Taylor, "Yemen's Counter-terrorism unit," *al-Jazeera*, September 17, 2008, accessed January 13, 2015, http://www.aljazeera.com/news/middleeast/2007/07/2008525185044972283.html

Chapter 2

Afghanistan

The Land of the Unconquerable

THE MODERN HISTORY OF AFGHANISTAN

The history of Afghanistan is turbulent; its people are resilient and its terrain is unforgiving. When the people and the land are combined, Afghanistan authenticates the title, "The Land of the Unconquerable." From a modern history perspective, Afghanistan was relinquished by the British on August 20, 1920, although it was never actually a colony. Exhausted and frustrated by the fierce Afghans, the British let go of the country and left it to its own machinations. What followed was a period of unsuccessful self-rule, which led to numerous executions, including the executions of government officials. In 1953, the impoverished Afghans turned to the Soviet Union for economic and military funding. What began as a request for assistance eventually led to a communist coup in April 1978, and later led to unwelcome tyrants when the Soviet Union invaded Afghanistan in December, 1979. A proxy war was then played out between the USSR and the United States on Afghanistan soil, and the country once again became a battleground.[1]

During the Soviet occupation, expert Afghanistan scholar, Barnett Rubin, states that six phenomena occurred in Afghanistan: (1) dependence of competing leaders on opposing flows of politically motivated military assistance; (2) dependence of the population for subsistence on politically motivated humanitarian aid; (3) destruction of the rural subsistence economy through counterinsurgency; (4) forced urbanization, including internal displacement to Afghan cities and the flight of millions of mainly rural refugees to camps and cities in Pakistan and Iran; (5) creation of refugee-warrior communities in Pakistan and Iran and of a region-wide Afghan diaspora and rapid monetization of the economy.[2]

These phenomenon in combination helped to create the right environment for the birth of the Taliban and Islamic mujahidin groups, such as al-Qa'ida[3] that are targeted by American drones today in Afghanistan. The rural population was dependent upon Soviet aid that went through so many channels that a lot of money was lost either through bribery, theft, or extortion in the process. The Taliban was the premier recipient of this "lost" funding. Weapons were resold at higher prices and were easy to transport in addition to drugs, mainly opium. Opium was easy to grow in most areas of Afghanistan and often provided work for young, landless men. The government had failed to maintain control over much of Afghanistan and warlords frequently filled in for the government, exacting monetary tribute in return for protection and subservience.

In addition to the Taliban, mujahidin groups grew tremendously during the Soviet occupation. Afghans who were previously unsupportive of mujahidin groups became much more receptive as a result of the Soviet occupation, which resulted in massive violence and oppression. Money from the Soviet Union, Pakistan, Saudi Arabia, the United States, the Muslim Brotherhood, and Salafis in Kuwait flowed into the country. The United States, in particular, was happy to fund Islamic extremists as long as those extremists resisted the communists, including Osama bin Laden and the early antecedents to al-Qa'ida.[4] The proxy war in Afghanistan lasted until February 1989, when the Soviets finally left.

Originally the Taliban began as teachers in madrassas that were composed of young men who had been orphaned during the Soviet occupation. The Taliban continued to grow, as they harnessed the anger and power of these devout Muslim men. The Taliban took over Kandahar in 1994; and in 1995, they took Herat and Jalalabad; and they took Kabul in 1996. In 1998, they massacred approximately 2000 people in Mazar-I-Sharif in revenge for the Taliban men that died in a failure to take the city previously. The Taliban also used local tribal ties to establish a network across the madrassas and to create solidarity across the country. Pakistan assisted the Taliban with military training and advisors to keep Iran from gaining control in the region. Because the Taliban controlled the roads, goods could travel safely throughout the country including opium, as the Taliban charged a 20 percent tax on opium traders. Opium is approximately 30–50 percent of the gross domestic product in Afghanistan. Afghanistan currently supplies 90 percent of the world's illegal opiates[5] and gets somewhere around 70 million to 400 million from opium production per year. Scholar Gretchen Peters[6] estimates that the Taliban earns as much as 500 million per year in cash and commodities from the drug trade. Other Taliban income comes from the *zakat* or tax charged by the mosques.[7,8]

According to academics, Brown and Rassler, the Taliban used the Haqqani Network[9] to act as a regional bridge to adapt its strategies to the local population and to infiltrate the tribes. The Haqqani Network is an Islamic fundamentalist network, created by Jalaluddin Haqqani (a good friend of Osama bin Laden) in 1973 who mostly acts within eastern Afghanistan and the FATA region. Although the Haqqani Network wants to reestablish the caliphate and views the United States as the infidel, typically, they do not carry out terrorist attacks. Since the Taliban's arrival in the early 1990s, the Taliban has had a deal with the Haqqanis to serve as a facilitator for Taliban relations so long as the Taliban let the Haqqani Network operate independently within their own territories.[10]

In the book *Talibanistan*, edited by Peter Bergen, the story of the Taliban in four different regions is expounded upon by various authors. It is important to state that the Taliban varies from region to region as drones target Taliban all over Afghanistan. The Taliban must not be viewed as a single entity in Afghanistan but instead as a heterogeneous group that is loosely allied throughout various regions in Afghanistan. Anthropologist James Ferguson relays the following quote from a British Special Air Services (SAS) officer in his book, *The Taliban*.

That is the thing that bothers me most, said the SAS officer. "In 2006 when the fighting started, we called everyone who resisted us 'Taliban.' But they really weren't necessarily. They were just the community's warrior class who had always defended their community against outsiders, and were bound to do so again. The 'Taliban' in that sense were an enemy of our own creation."[11]

Journalist Tom Coglan adds that there are the real Taliban who are opposed to the government of Afghanistan and the establishment of government backed by Western forces and there are also Taliban jihadis who only want to be martyred in the process of making their jihad.[12]

The other group that is frequently targeted with American drones in Afghanistan is al-Qa'ida. In contrast to the Taliban, al Qa'ida came out of the 1979 revolution in Afghanistan and Osama bin Laden moved the headquarters of al-Qa'ida from Afghanistan to Peshwar, Pakistan, and then to Sudan. In the mid-1990s al-Qa'ida moved back to Afghanistan. Similar to their relationship with the Taliban, the Haqqani Network also assists al-Qa'ida. Brown and Rassler state, "For the past twenty years, the Haqqani Network has consistently been al-Qa'ida's most important local partner and facilitator in the region, contributing to the operational development of Bin Laden's group in many ways. The reason why the Haqqani network's role has not been fully appreciated is because the assistance it provides to al-Qa'ida and others is locally oriented and most often transpires in the background."[13] Haqqani also allows al-Qa'ida and other jihadists to use their training camps, even stating there "is no distinction between us . . . we are all one."[14] Haqqani also provides safe haven to al-Qa'ida members and perhaps has even funded al-Qa'ida training camps.

The drone war in Afghanistan initially began on September 11, 2001, when three hijacked planes crashed into the Pentagon and the twin towers of the World Trade Center. Another plane, headed toward the White House crashed in a field in Pennsylvania. Mastermind of the attack was the founder of al-Qa'ida, Osama bin Laden. After September 11, 2001, during the American invasion, the Taliban and al-Qa'ida became the primary enemies in the War on Terror while the Haqqani Network was rarely mentioned. The Taliban became a target as it was responsible for allowing al-Qa'ida to train in Afghanistan, in addition to some ideological connections and similarities.

The War on Terror quickly ensued in an effort to destroy all terrorists in the world and to defend the American homeland. The first phase of the War on Terrorism in Afghanistan lasted from October 2001 to February 2002. During this phase, the United States launched drones against Taliban positions. The Taliban eventually surrendered and the operations in Tora Bora and Shahikot sent the last of al-Qa'ida into Pakistan. The December 2001 Bonn Accords provided a peacekeeping force, and The International Security Assistance Force (ISAF) was mandated by UN Security Council Resolution 1386.

However, after the initial invasion, the Taliban resurfaced in Afghanistan. United States' Ambassador to Afghanistan, James Dobbins,[15] attributes the renewed civil war in Afghanistan after the initial invasion to two causes: (1) the United States, the Karzai Administration, and the rest of the international community failed to take action against the Taliban after the lull of the initial invasion in 2001 to increase government services and security into the rural areas. The period after the initial invasion was the best time to physically destroy the Taliban. (2) The second problem was the regional consensus that was reached at Petersberg, concerning the confidence in President Karzai of Afghanistan, which had substantially degraded. Dobbins also states that the Bush Administration failed to follow through on the nation-building in Afghanistan after the initial invasion.

In August 2003, NATO took control of ISAF command. The second phase of military operations occurred from March 2002 to the end of 2005. The goals of this phase were to decrease ethnic regional tensions and to recreate and train an Afghanistan National Army. Small pockets of Taliban were still creating problems in the Southern provinces.

The third phase of military operations from 2006 to mid-2009 consisted of a consolidation of all security forces. In 2007, the United States took command of all ISAF forces under General Stanley McCrystal as a majority of the forces were American anyway. In response, the Taliban began activating forces in the North, and outside of Kabul and continued to grow. Kidnapped journalist, David Rohde, attests to the resurgence in Taliban numbers in late 2008 when he spent over seven months in captivity in Afghanistan. Rohde states that he was astonished by "a Taliban mini-state that flourished openly and with impunity. The Taliban government that had supposedly been eliminated by the 2001 invasion of Afghanistan was alive and thriving."[16]

The last phase of operations began in mid-2009 with a troop surge of around 30,000 American soldiers. This was also the same time period that drones became more common in Afghanistan particularly as being able to watch "patterns of life" or being used as surveillance technology.[17] General Stanley McChrystal was replaced with General David Petraeus in July 2010 after a scandal in which the General criticized the Obama Administration and his close aides. Petraeus received an additional 20,000 troops over the Spring and Summer of 2010. The National Security Strategy of the United States in 2010 stated:

> In Afghanistan, we must deny al-Qaida a safe haven, deny the Taliban the ability to overthrow the government, and strengthen the capacity of Afghanistan's security forces and government so that they can take lead responsibility for Afghanistan's future. . . . First, our military and ISAF partners are targeting the insurgency, working to secure key population centers, and increasing efforts to train Afghan security forces. . . . Second, we will continue to work . . . to improve accountable and effective governance . . . focusing assistance on supporting the President of Afghanistan and those ministries, governors, and local leaders who combat corruption and deliver for the people. . . . This will support our long-term commitment to . . . a strong, stable, and prosperous Afghanistan.[18]

However, Taliban violence continued to escalate and climaxed in 2011 as the Taliban became efficient in infiltrating Afghan forces and attacking them from the inside.[19] From August to November 2010, in response, 3064 bombs and missiles were released in drone strikes, more than twice the number of bombs dropped during the same period in 2009.[20]

Chris Woods in his book, *Sudden Justice*,[21] makes the excellent point that the increases in drone strikes in Afghanistan occurred for two good reasons. President Obama issued Executive Order 13491 two days after he was inaugurated, outlawing the use of torture (enhanced interrogation techniques) and closing down ghost prisons where people of "questionable intent" were kept. In addition, JSOC and the CIA were required to use the *US Army Field Manual* disallowing them from using their own tactics and strategies. Non-profits and watchdogs like The Red Cross could visit any United States facilities at any time. In essence, what occurred was that President Obama halted torture and detainment policies and instead created a "kill first, question later policy." Organizations such Amnesty International or Human Rights Watch could not bring a case against the United States' military as effectively if the person of interest was already dead.

Since the Americans left Afghanistan in 2013, save for the military bases, drones are being used to kill terrorist elements throughout the country "with one in ten Afghan airstrikes in 2010 now by remotely piloted aircraft."[22] This number has grown drastically since 2010 since drones are easier to use and most of the troops have left. Operation Freedom Sentinel, which replaced Operation Enduring Freedom (in place since 2001) began at the end of 2014. Drone strikes in Afghanistan are the highest in the world. Aimal Faizi, the former spokesperson for former President Hamid Karzai, asks why Afghanistan has "become the most heavily drone-bombed country in the world?" As can be seen in Figure 2.1: *American Drone Strikes and Deaths in Afghanistan 2015*, the number of drone strikes has grown tremendously just in 2015 with over 1,517 confirmed deaths and 140 drone strikes according to *The Bureau of Investigative Journalism*.[23] Although this data includes the maximum deaths[24] of people killed by drones and air strikes are not included, there are still several strikes which have an unknown number of deaths that are not incorporated below. This number of drone strikes is more than Pakistan, which had approximately fifteen drone strikes in 2015 and is frequently given the title of the most heavily drone-bombed country in the world. Approximately 62–85 people have died from these drone strikes in Pakistan alone in 2015.[25]

Month	Number of Drone Strikes	Possible Deaths-Highest Estimate
January	6	49
February	4	44
March	8	49
April	3	10
May	10	133
June	11	239
July	20	347
August	13	148
September	18	112
October	22	264
November	10	100
December	15	22

Figure 2.1 American Dronke Strikes and Deaths in Afghanistan 2015. *Source*: Serle, Jack, "Get the data: A list of US air and drone strikes, Afghanistan 2015," *The Bureau of Investigative Journalism*, February 12, 2015, accessed August 13, 2015, https://www.thebureauinvestigates.com/2015/02/12/us-drone-war-afghanistan-list-american-air-strikes-2015/#AFG066.

What is particularly evident in Figure 2.1 is the number of deaths attributed to drone strikes in Afghanistan. In 2015, from January to July, in Afghanistan alone, there was an average of 14.05 deaths per drone strike including the "terrorists" that are targeted. Considering that drone strikes are usually used for targeted killing of one or two people, 14.05 deaths per strike belies a lack of concern for human life. It is highly unlikely that the United States Air Force is supposed to target 347 people per month, similar to what occurred in July 2015. These numbers are alarming in that there is not enough attention being paid to civilian life or collateral damage. The United States started having the Afghan Air Force conduct their own airstrikes in August 2015 with higher civilian casualties than the American airstrikes and drone strikes. The numbers for the entire year of 2015 for American drone strikes decreased for a total of 10.83 deaths per drone strike.

According to the military's collateral damage methodology protocols, planned collateral damage should be less than 10 percent, this means there should be one civilian killed for every ten drone strikes or 0.1 persons killed per drone strike. This would also mean that the military must be sure that there is a less than 10 percent chance that civilians will be hurt in a drone strike. The United States was killing 10.83 people per drone strike in 2015 in Afghanistan.[26]

THE RESULTS OF AMERICAN DRONE WARFARE IN AFGHANISTAN

The following section explains from several perspectives, including both qualitative and quantitative research methodologies, the effects of drone warfare on Afghanistan. The results are not positive in that it is evident that there are several problems with the drone policy in Afghanistan.

What Exactly is a "Person of Interest" and Who is in Charge Here?

The question of defining who is a "terrorist" and who is in charge of military operations have been frequent problems in Afghanistan. Chris Woods talks about the difficulties that the military had with who was considered "a terrorist" in Afghanistan and what could be done with them in the early stages of targeted killing in Afghanistan. Although persons had been questioned and nothing pertinent had been found, arguments would be found to continue with the assassination. "There then followed this conversation. 'OK, so how long do we need to wait [after] he's been released from custody before it would be appropriate to kill him.' Obviously, the easiest place to kill him was as

soon as we've let him out of the front gate, but that would have been a bit unsporting."[27]
In my own visit to Holloman Air Force Base, the major training center for Air Force drone pilots on September 10, 2015, I was given a similar impression. Remotely Piloted Aircraft pilots or drone pilots in the United States' Air Force often follow a target for hours at a time. There is a lot of surveillance that goes into targeting a "terrorist." However, the reasons for killing that person are not necessarily pertinent. In fact, the pilots are not even aware of why they are told to kill a person. It seems as though once a person has been deemed "dangerous" by the President, CIA, or Central Command (CENTCOM) regardless of the urgency of the threat, that person is disposable. Interrogating the target does not seem to be a concern in that anyone who is considered a "target" or worth targeting is killed. Around 2001, pilots were given more freedom to "ease" terrorist targeting restrictions by Secretary Donald Rumsfield. This means that instead of going through the extended chain of command, less people were involved in the decision-making process.[28] However, this process is still mysterious as the American Government has not published its protocol on the matter.

Woods also expounds upon all the numerous problems with drone strikes in Afghanistan concerning military personnel cooperation through personal interviews. US Special Forces would do what they wanted and ISAF Forces were usually left to clean up the mess. "You would get a phone call at half midnight and be given four grid references and told. . . . 'We're taking over your battle space for the next three hours, move all of your units out. We will give you the battle space back at the end.' And then you would say 'What are you going to do?' 'We can't tell you. You will find out later.'" This statement was made by former British Army Intelligence Officer Mike Martin.[29] Martin also stated in a personal interview with Chris Woods.

> They were killing people and they didn't even know their name let alone their tribe. They just picked them up from signals intelligence. And so we repeatedly tried to say there's a wider context out here. In the end, they introduced something called a "No Strike List" . . . but getting people put on that was like turning wine into water, because the Special Forces were incredibly powerful and they could basically do whatever the fuck they wanted.[30]

Besides the lack of cooperation on the American side and the so-called "trigger finger" issue with CENTCOM, civilian deaths are also a concern in Afghanistan.

Civilian Deaths and More: Death by the Numbers

Unlike several of the other countries that are examined within this book, the civilian casualty rate caused by American drones in Afghanistan is unknown before 2009. The main reason that drone casualty rates are unknown in Afghanistan is that no one was studying drone deaths in Afghanistan. Unlike in Pakistan, Somalia, or Yemen, think tanks like *The Bureau of Investigative Journalism* were not concentrating on Afghanistan. Only recently in 2015 has *The Bureau of Investigative Journalism* began to catalog drone strikes in Afghanistan. As recently mentioned, Afghanistan experiences the most drone strikes in the world.

Another reason for this lack of data collection is that the United States' Air Force stopped publishing data on drone strikes in 2013 in Afghanistan and this data became classified.[31] Analysts are unsure as to what causes a death when it occurs if no one actually admits to a drone strike and no one is there to catalog it. The American military is not responsible for keeping track of the people they kill.

The Bureau of Investigative Journalism[32] states on their website that the United Nations Assistance Mission in Afghanistan (UNAMA) shared the following information privately with them. In 2009, UNAMA put the number of drone deaths in Afghanistan at zero, although this is hard to believe. In 2012, civilian drone deaths were estimated as sixteen. Estimates in 2013 put the civilian death toll as high as forty-five. In the first half of 2014, there were eight casualties. As already mentioned, 1517 people in Afghanistan were killed in drone strikes in 2015. Figure 2.2 is a chart taken from the UNAMA Report.[33] The number of aerial casualties and strikes are given in Figure 2.2: *Civilian Deaths and Injuries from Aerial Operations, Afghanistan 2009–2014*. It is unclear as to how many of these "aerial operations" are drone strikes (drone strikes are included in the term) but this does give the

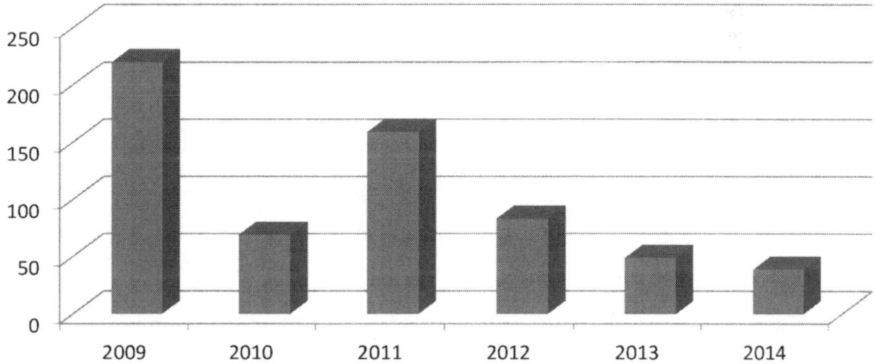

Figure 2.2 Civilian Deaths and Injuries from Aerial Operations, Afghanistan 2009–2014. *Source*: United Nations Assistance Mission in Afghanistan (UNAMA), "Afghanistan Midyear Report 2014, Protection of Civilians in Armed Conflict," July 14, 2015, accessed August 3, 2015.

reader some idea of how high drone casualties could be. UNAMA states that there could be more aerial operations than listed below as not every strike is reported. What is interesting is that civilian deaths and injuries have decreased over the years, although in 2015, they have drastically increased. UNAMA attributes this to ISAF taking greater care to stop civilian casualties albeit recently civilian deaths have increased exponentially.

Mistakes do occur and unfortunately, they occur more often than not. Although these mistakes are easy to find and they are often seen in the newspapers, American drone warfare in Afghanistan has had its fair share of mishaps. One such event occurred in February 2010, when three vehicles in Uruzgan Province were attacked by helicopters after extensive Predator surveillance (approximately three hours). Around twenty-three men were killed and twelve others were injured including one woman and three children. It was later revealed that the people were actually farming families who happened to be in the wrong place at the wrong time. Woods[34] talks about pilots being excited to kill the "squirters" or those people who run after the first hit has occurred. Blame in this situation seemed to be thrown everywhere but 175,000 was paid in compensation by Afghan forces to the families and personnel were disciplined.

Former President, Hamid Karzai, pleaded repeatedly with the United States to stop killing civilians with drone strikes. Karzai's own cousin Haji Yar Mohammad was killed in an American military night raid in the Khandahar region in early March 2011.[35] Karzai even refused to sign a proposed ten-year security agreement with the United States unless civilian casualties decreased. "Karzai responded to the Kunar attack by accusing the United States of recklessness and callous disregard for innocent Afghans."[36] To add to this problematic situation, no high-profile terrorists have been killed in Afghanistan. To date, the deaths consist of rank and file, al-Qa'ida and Taliban members.

Most civilians do not receive compensation for their losses after a drone attack. In a report from the *Center from Civilians in Conflict* concerning research conducted in 2012, Afghans rarely receive compensation from a guilty party for the death of a loved one. Even though international forces and Afghan National Security Forces have taken steps to provide compensation, several problems exist. The system is extremely slow and it often takes months to get any money, which is often a small amount. The filing system is also extremely complicated to use and arduous. Officials often expect a bribe to do anything for people who are already financially burdened.

Women who lose male protectors are initially at a tremendous disadvantage. When a woman loses a husband or son, she is bound by society to ask for her brother-in-law to file the paperwork with the proper authorities. If there is any compensation, the brother-in-law will further take responsibility in doling out the money, thus making women more financially dependent on men in an already patriarchal society. Many women who do not have family are forced to beg as their gender keeps them from filing the paperwork in the first place.[37]

"They are Terrified of the Drones"

Afghanistan, an important country in the War on Terrorism for terrorist recruitment and training, has some mixed messages concerning the effectiveness of drones and drone warfare. The Taliban and al-Qa'ida are certainly fearful of drones. David Rohde was kidnapped and held for seven months in North and South Warziristan. Rohde states that his Taliban guards were terrified of the drone strikes and even tortured a local farmer who they thought had given the location of the camp from a GPS that was located in his tire. Rohde who is a columnist for *Reuters* states, "He said he was innocent, but a group of foreign and Pakistani militants tortured him. First, they chopped off his leg and then disemboweled him, at which point he "confessed" to being a spy. I was told that they decapitated him and hung his body in the market in Makeen as a warning to other villagers."[38] Rohde believes that the drones had found their mark but Taliban leaders were never killed. Rohde was also concerned about the civilian deaths that accompanied the targeted killing.

As can be seen in Figure 2.3: *Afghanistan's Biggest Problems 2014*,[39] public opinion, collected by The Asia Foundation highlights the daily fears that Afghanistan people face concerning terrorism and violence. Thirty-four percent of Afghans in 2014, or over one-third throughout the entire country, were

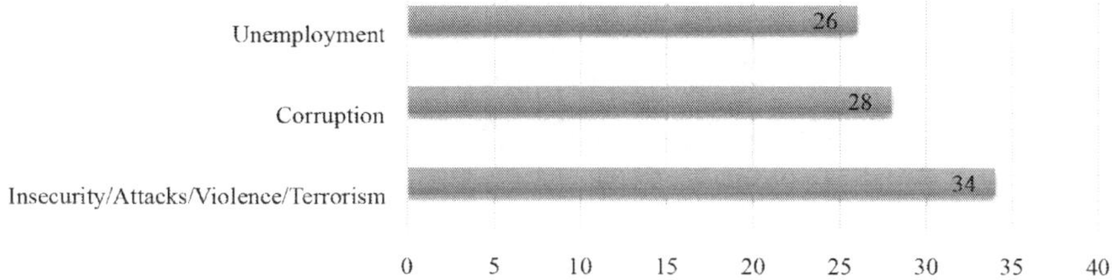

Figure 2.3 Afghanistan's Biggest Problems 2014. *Source*: The Asia Foundation, "Afghanistan in 2014: A Survey of the Afghan People," November 18, 2014, accessed November 2, 2015, http://asiafoundation.org/resources/pdfs/Afghanistanin2014final.pdf.

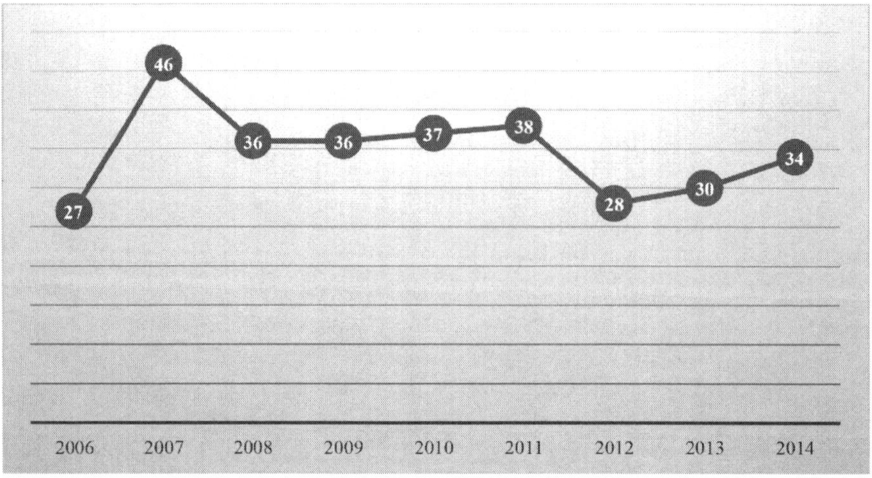

Figure 2.4 Afghanistan's Biggest Problem by Year—Insecurity, Attacks, Violence, and Terrorism. *Source*: The Asia Foundation, "Afghanistan in 2014: A Survey of the Afghan People," November 18, 2014, accessed November 2, 2015, http://asiafoundation.org/resources/pdfs/Afghanistanin2014final.pdf.

afraid of violence. In 2014, 65.4 percent of Afghans report always, often, or sometimes fearing for their safety or security or that of their family. This fear for safety has increased 59 percent since 2013 and has continued to grow since 2006. Provinces reporting the highest levels of fear for personal safety are Faryab (92.1%), Wardak (89.2%), Farah (87.5%), and Kunduz (87.5%). Seventy-five percent of Afghans say they would be afraid when traveling within Afghanistan.[40]

This problem is further expounded upon in Figure 2.4: *Afghanistan's Biggest Problems by Year—Insecurity, Attacks, Violence, and Terrorism*.[41] The biggest problem for the people in Afghanistan is violence, whether American drone strikes, terrorist attacks, etc. These Afghan people live in constant fear, which has been occurring for over four decades but has recently increased again. Since 2006, there is not a tremendous amount of change in the fear of citizens in their daily lives.

Anti-Americanism and Taliban/al-Qa'ida Support

The level of hatred for the United States and President Obama is evident and has definitely increased since the initial invasion in Afghanistan. Rohde stated that during his captivity with the Taliban, "I watched one Taliban commander spit at a picture in the Pakistani newspaper of Barack and Michelle Obama dancing at an inaugural ball."[42] The Taliban viewed the United States as cowardly and wondered why the country was afraid to fight a ground war. "As a result, my Afghani Taliban guards vowed to carry out revenge attacks in the United States. . . . Overall these drone strikes seemed to create a stalemate. They limited operations and killed some senior militants. However, they certainly did not stop these groups from being active in tribal areas."[43] Aimal Faizi, the former spokesperson for President Karzai and Afghan journalist stated, "Beyond the civilian casualties caused by these strikes, drones also fuel terrorism, increase anti-government sentiment and, as a result, increase recruitment opportunities for the armed opposition in Afghanistan."[44]

Quotes expressing anti-Americanism are plentiful. Another Afghan citizen stated after an airstrike occurred, which hit his brother's house, "I can give you lots of money and then kill seven Americans . . . human life is not for sale, Nothing can bring back my family."[45] This citizen received compensation for his losses but that compensation did not make up for the loss of his family. Another citizen stated, "We don't want international forces here. Foreigners should leave Afghanistan. . . . They shouldn't kill innocent people. . . . We were a very happy family before."[46]

Although it is difficult to assess whether al-Qa'ida or the Taliban have grown or decreased in numbers as a result of drone strikes, as terrorist organizations are secretive by nature, it is possible to assess public support for the groups. In a survey of ninety-two districts in Afghanistan, not one district supported the Afghanistan Government in December 2009. The number of neutral districts was forty-four and the number of districts who sympathized with the insurgency climbed from 33 to 48 in March 2010.[47]

The Resurgence of the Taliban in Kandahar and Other Areas

Although early reports pointed to the destruction of the Taliban, Anand Gopal argues that the Taliban regrouped in Pakistan and have continued to grow in the Kandahar region and Kandahar City since 2009. "The resurgence of the Taliban in Kandahar was not preordained simply because the province was their "spiritual home." Nor was it merely a result of a lack of troop presence in the early years. Rather it was due to specific policies pursued by the Kandahar government and its American backers."[48]

Changes in Operation and Tactics

The use of drones has caused changes in tactics for the Taliban in Afghanistan. According to Gopal, the Taliban has changed to using small-scale tactics from large group attacks because small-scale tactics are easier to use when under constant drone surveillance and the pressure of strikes. The number of suicide bombings since 2006 has increased tremendously as can

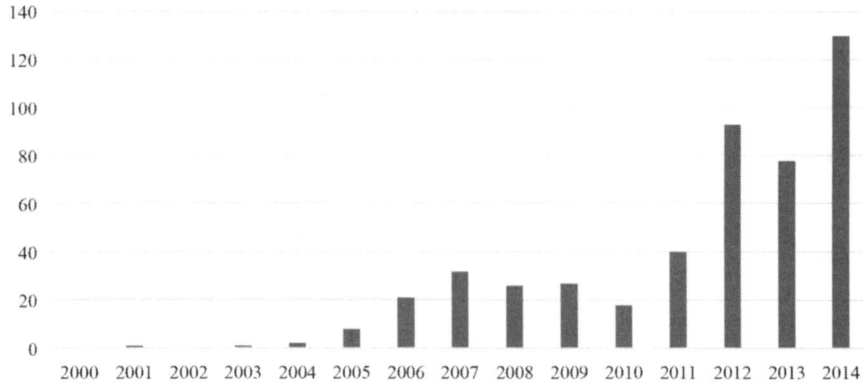

Figure 2.5 Taliban Suicide Attacks 2000–2010 Afghanistan. *Source*: Pakistan Body Count.org. "Pakistan Body Count." 2015. Accessed September 16, 2015, http://pakistanbodycount.org/suicide_bombing.

be seen in Figure 2.5: *Taliban Suicide Attacks 2000–2010 Afghanistan.* The roadside bomb has also been employed more against foreign troops. In addition, the Taliban has begun assassinating anyone who collaborates with the legitimate Afghanistan Government under Karzai including tribal elders and government officials. Schools also have become more common targets. Many Afghans have joined the Taliban so as not to become a victim of the Taliban. One Afghanistan citizen stated that "Suicide bombers prefer to die at dawn, or early in the morning after praying and first ablutions, so we all knew that if we hadn't heard explosions by eleven or midday, there was a good chance we'd get through the day without too much harm."[49]

In late 2009, the Taliban escalated both assassinations and the push into Kandahar City. This was caused by the American troop surge in the area and the growth of the Taliban in the Kandahar region. Gopal argues that the Special Forces campaign, which included the use of drones to assassinate Taliban members, was not successful. Instead it only increased the violence in the Kandahar region as security incidents per week increased, so did Taliban attacks in the region of Kandahar. Dead Taliban commanders were immediately replaced by younger men.[50]

Another Afghanistan expert, Martine van Bijlert, agrees that the killing of Taliban leadership is not effective. Attempts on Taliban leadership have instead lead to tension between locals and international forces. "In response to the heavy losses inflicted the overwhelming airpower of the international military, the Taliban shifted early on from large-scale frontal attacks to guerilla operations and asymmetrical warfare: ambushes, assassinations, small-scale attacks, IEDs, suicide bombings. The coalition's targeting of high-and midlevel commanders, which has been going on for years and has taken out a large number of prominent fighters, kept the movement in flux but had relatively little impact on its operational capabilities."[51] Although, van Bijlert does state that moral was weakened by the targeted assassinations.

Quantitative Analysis and Results

Jaeger and Siddique found that "there is no significant impact of drone strikes on Taliban and Al-Qa'ida attacks in Afghanistan"[52] through vector auto-regression. In addition they found that a terrorist attack is more likely to occur after a drone strike, "A terrorist attack in Afghanistan is 8.8 percent more likely five days after a successful drone strike."[53]

When data were collected and analyzed for this chapter through linear regression,[54] it was also found that there was no significant impact of either drones on terrorist attacks or terrorists attacks on drones. The author can safely say that when the United States uses drone strikes to kill terrorists in Afghanistan, the number of terrorist attacks in Afghanistan do not decrease. It can also be said that when a terrorist attack occurs, it does not lead to an increase or decrease in drone attacks. This relationship is the same for suicide bombings. Although suicide bombings have increased, there is no relationship between drone attacks and suicide bombings; neither one leads to any effect on the other. Jaeger and Siddique were likely correct when they found that "A terrorist attack in Afghanistan is 8.8 percent more likely five days after a successful drone strike."[55] Figure 2.6: *American Drone Strikes and Terrorist Attacks, Afghanistan January 2007–December 2013* attests to the fact that there does not seem to be much of a relationship between drone attacks and terrorist attacks. According to the Afghanistan Taliban website, Islamic Emirate of Afghanistan, the Taliban was the most interested in getting American troops to leave Afghanistan, rather than a tit for tat cycle.

Maintaining American troops in Afghanistan can in no way slow down the rapid process of our Jihad and struggle. The insistence upon policy of war by American officials will only further aggravate the sensitivities of our nation and the region. When attacks begin to intensify against the American invaders, their casualties begin rising and the cost of this futile and unwinnable war in Afghanistan increases, she will willfully begin changing her tyrannical roadmap. We call upon our Mujahideen to intensify their attacks against American targets and to revise, escalate, and quicken their plans against their movements, bases, and affiliated organs.

Until the United States leaves, the Taliban will continue to use every weapon at their disposal to attack American forces.[56]

Figure 2.6: *American Drone Strikes and Terrorist Attacks, Afghanistan January 2007–December 2013* shows American drone strikes and terrorist

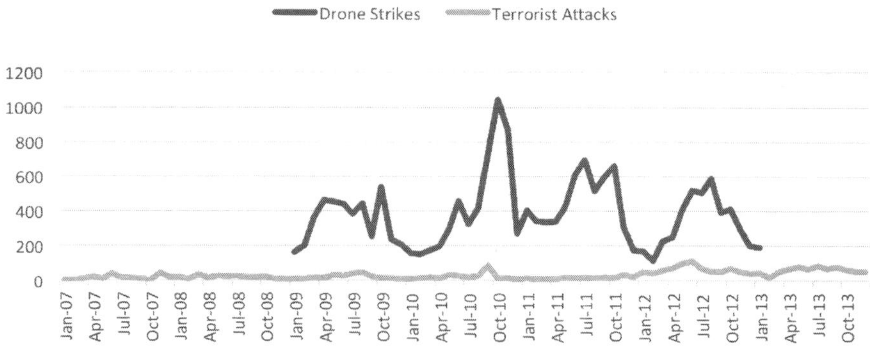

Figure 2.6 American Drone Strikes and Terrorist Attacks, Afghanistan January 2007–December 2013. *Source*: For 2009–2012 (Data below was pulled by the US Air Force in 2015): Noah Shachtman, "Military Stats Reveal Epicenter of US Drone War," Wired.com, November 9, 2012, accessed February 23, 2016, http://www.wired.com/2012/11/drones-afghan-air-war/; American Drone strikes for 2013 and 2014 in Afghanistan are not available because they are still classified by the United States Air Force. For 2015: Jack Serle and Payenda Sargand, "Get the data: A list of US air and drone strikes, Afghanistan 2015," *The Bureau of Investigative Journalism*, February 12, 2015, accessed August 13, 2015, https://www.thebureauinvestigates.com/2015/02/12/us-drone-war-afghanistan-list-american-air-strikes-2015/.

attacks committed by mostly Taliban and some al-Qa'ida. It is evident that American drone strikes have little effect on the number of terrorist attacks that are committed. A report from the Joint and Coalition Operational Analysis found that "Drone strikes in Afghanistan were seen to have close to the same number of civilian casualties per incident as manned aircraft, and were an order of magnitude more likely to result in civilian casualties per engagement."[57] The increase in civilian casualties, although it leads to a decrease in American military injuries and fatalities, is not the optimum desired result.

CONCLUSION

In conclusion there are several troubling components of the American drone policy in Afghanistan, the primary being that it does not seem to be working. Drone strikes in Afghanistan lead to civilian deaths, property destruction, anti-Americanism, al-Qa'ida, and Taliban support, in addition to many other issues. The most troubling problem is that drone strikes do not lead to less terrorist attacks or suicide bombings in Afghanistan. In fact, it has absolutely no effect on them. What has occurred instead is an Afghanistan population that is riddled by drones on a regular basis and many Afghans have blamed the United States.

One thing that should have been included in this chapter is the number of terrorists that the drone strikes have killed. However, since there is little background on who the targets are, it is often impossible to even identify them as a Taliban member or a terrorist. Considering that Haqqani, Taliban, and al-Qa'ida work together, it is sometimes suspect as to what group the terrorist even belongs to or in some cases if they were just at the wrong place at the wrong time.

Another issue to consider is the triangular relationship between Pakistan, Afghanistan, and the United States. It was difficult to compile data and sources for this chapter at times as Pakistan receives so much more attention than Afghanistan. Because the two countries border one another, Taliban and al-Qa'ida targets often get lost and found in the mountainous region between the two countries. Therefore there must be extensive cooperation between the Afghan army, the Pakistan army, and the American military components to target or find terrorists. Chris Woods stated that General Kayani, Pakistan's military chief, stated in a report handed to President Obama. "Mr. President, this is my analysis of where and why you're going wrong in Afghanistan. And this will also explain why I am afraid that I cannot meet your expectations in what you want me to do."[58] Pakistan has lost thousands of soldiers assisting the United States in the War on Terror yet Osama bin Laden was found hiding in Abbottabad, Pakistan.

In the effort to win in Afghanistan, Pakistan must be inherently included in the planning. In the last chapter of his book *Operation Dark Heart*, Lt. Col. Anthony Shaffer provides his advice on how to win the War on Terror in Afghanistan. His first piece of advice is to "Create an Internal Operational Policy Framework—The Internal Defense and Development (IDAD) Program" to train, advise, and equip the Afghan army, Air Force, and National Police. The United States has somewhat fulfilled this piece of advice. (2) The next thing Shaffer advises is to control the border between Pakistan and Afghanistan. The border is porous and most Afghanis and Pakistanis cross the border at will, allowing other problems to cross the border at will also. A solid secure border would keep the terrorists in the same area, not allowing them to escape. (3) The third recommendation is the reduction in all violence in the lives of citizens, including government violence. Citizens must be allowed to live in peace and the collateral damage caused by drones must end. It is clear from the evidence provided in this chapter that for most generations of Afghans that peace has been an elusive concept. (4) The fourth thing that must be done is to change the analytical approach that the United States uses in Afghanistan and Pakistan. "We have to become involved in helping to shape and improve the message of the true Muslim faith."[59] It must be understood that power and religion are synonymous and a better understanding of Islam and the

Afghanistan culture must occur although the United States cannot be held responsible for teaching or understanding true Islam. (5) The fifth recommendation is that the United States must change its logistical footprint and refrain from leaving a mini-United States in Afghanistan. Since the United States has not left Afghanistan completely, this is still a possible consideration. (6) The sixth suggestion is that the United States needs to follow the money trail of the Taliban to defeat the Taliban. Money funds the Taliban and the Americans have no concept of identifying that money trail, particularly regarding the opium that the Taliban sells. (7) The seventh suggestion is that there must be a real peace process and Pakistan cannot be used as a safe haven for Taliban. (8) Eighth, the United States must address the real cause of regional instability as the Pakistani–Indian Cold War. Pakistan is concerned about an Indian regional hegemony and acts accordingly. The United States must take this Pakistani cause into account. (9) Lastly, the price of failure in Afghanistan is too high to let the possibility occur. If the United States does not leave Afghanistan with a sovereign and strong government, then it is likely that the United States will end up in Afghanistan again.[60]

It is still possible to change the course of American targeted killing in Afghanistan. As will be stated throughout this book, targeted killing in the author's opinion, regardless of the country, must be used rarely and effectively. Fifteen people cannot be killed in every drone strike; this leads to terrorist group empathy, sympathy, and easy recruitment. In addition, it is just plain sloppy. If the United States is to continue targeted killing on a limited basis, it must be clean, justified, and the United States must take public responsibility for those people that it kills. The lack of public responsibility allows the United States to be blamed for actions it may not have taken, but instead may have been taken by countries such as England who are also using targeted killing in Afghanistan. There must be oversight if the United States is to continue its targeted killing policy in Afghanistan. The president of the United States cannot be the sole decision-maker concerning who is targeted by American drones in Afghanistan.

NOTES

1. For an excellent account of this proxy war, see Steve Coll, *Ghost Wars: The Secret History of the CIA, Afghanistan, and bin Laden, from the Soviet Invasion to September 10, 2001* (New York: Penguin Books, 2005).

2. Barnett R. Rubin, *Afghanistan from the Cold War to the War on Terror* (New York: Oxford University Press, 2013), 55.

3. For an excellent examination of Afghanistan mujahidin groups, see Rubin, *Afghanistan from the Cold War to the War on Terror* "Chapter 3 Arab Islamists in Afghanistan."

4. Rubin, *Afghanistan from the Cold War to the War on Terror.*

5. Fariba Nawa, "Women and the Drug Trade in Afghanistan," in *Under the Drones*, eds. Shahzad Bashir and Robert D. Crews (Cambridge: Harvard University Press, 2012), 236–256.

6. Gretchen S. Peters, "The Taliban and the Opium Trade," in *Decoding the New Taliban, Insights from the Afghan Field,* ed. Antonio Giustozzi (New York: Columbia University Press, 2010), 19.

7. Rubin, *Afghanistan from the Cold War to the War on Terror,* 63.

8. Rosemarie Skaine, *Women of Afghanistan in the Post-Taliban Era, How Lives have Changed and Where They Stand Today* (Jefferson, NC: McFarland & Company, Inc., Publishers, 2008). Among numerous human rights abuses, the Taliban has greatly oppressed women during their tenure. Rosemarie Skaine's book *Women of Afghanistan in the Post-Taliban Era* elaborates on the human rights violations that women endured under the Taliban such as restriction of movement (confinement to the home), lack of medical care, and the prohibition of female education. Women are also not allowed to work, including those women that have been widowed. The Taliban would also assassinate women who were progressive or supported human rights within Afghan society. Besides misogyny, the Taliban uses violence as a way to control the population. Citizens are not really safe in this system and the Taliban and other terrorists constantly abuse the civilian population.

Please see Neamatollah Nojumi, Dyan Maurana, and Elizabeth Stites, *After the Taliban, Life and Security in Rural Afghanistan* (Lanham: Rowman & Littlefield Publishers, Inc.), 2009. For more information on life for rural Afghan people, see Peter Marsden, *The Taliban, War and Religion in Afghanistan* (New York: Zed Books Ltd.), 2002. For earlier information on women and their treatment by the Taliban.

9. Thomas Rutig, "Loya Paktia's Insurgency (I) The Haqqani Network As An Autonomous Entity," in *Decoding the New Taliban, Insights from the Afghan Field,* ed. Antonio Giustozzi (New York: Columbia University Press, 2010), 57–100.

10. Vahid Brown and Don Rassler, *Fountainhead of Jihad, the Haqqani Nexus, 1973–2012* (New York: Columbia University Press, 2013).

11. James Ferguson, *Taliban: The True Story of the World's Most Feared Guerilla Fighters* (London: Transworld Publishers 2010), 135.

12. Tom Coglan, "The Taliban in Helmand," in *Decoding the New Taliban, Insights from the Afghan Field,* ed. Antonio Giustozzi (New York: Columbia University Press, 2010), 132.

13. Vahid Brown and Don Rassler, *Fountainhead of Jihad, the Haqqani Nexus, 1973–2012* (New York: Columbia University Press, 2013), 185.

14. As cited in Vahid Brown and Don Rassler, *Fountainhead of Jihad, the Haqqani Nexus, 1973–2012* (New York: Columbia University Press, 2013), 191.

15. James F. Dobbins, *After the Taliban* (Washington D.C.: Potomac Books, Inc., 2008), 163–164.

16. David Rohde, "7 Months, 10 Days in Captivity," *The New York Times*, October 18, 2009, accessed September 14, 2015, http://www.nytimes.com/2009/10/18/world/asia/18hostage.html?pagewanted=all&_r=0.

17. Michael Waltz, "'Bring on the Magic,' Using Drones in Afghanistan," in *Drone Wars*, eds. Peter L. Bergen and Daniel Rothenberg (Cambridge: Cambridge University Press, 2015), 209.

18. Lewis G. Irwin, *Disjointed Ways, Disunified Means: Learning From America's Struggle to Build an Afghan Nation* (Carlisle, PA: Strategic Studies Institute, 2012), 1.

19. Michael Semple, "Afghanistan: The First Theater of War," in *Assessing the War on Terror*, ed. Mohammed Ayob and Etga Ugur (Boulder: Lynne Rienner Publishers, 2013), 47–70.

20. Chris Woods, *Sudden Justice, America's Secret Drone Wars* (New York: Oxford University Press, 2013), 218–219.

21. Chris Woods, *Sudden Justice, America's Secret Drone Wars* (New York: Oxford University Press, 2013).

22. Chris Woods, *Sudden Justice, America's Secret Drone Wars* (New York: Oxford University Press, 2013), 219.

23. The data for *The Bureau of Investigative Journalism* includes both air strikes and drone strikes. The author only included the drone strikes in these calculations.

24. Maximum deaths are used as opposed to minimum deaths, as these are typically reported by the local people who are familiar with and perhaps were at the scene of the strike. In addition, maximum deaths include people who were hit and later died from their injuries.

25. Jack Serle, "Obama 2015 Pakistan Drone Strikes." *The Bureau of Investigative Journalism*, January 5, 2015, accessed September 16, 2015, https://www.thebureauinvestigates.com/2015/01/05/obama-2015-pakistan-drone-strikes/.

26. Chairman, Joint Chiefs of Staff, "No Strike and the Collateral Damage Estimation Methodology," Chairman of the Joint Chiefs of Staff Instruction, February 13, 2009, Accessed August 21, 2015, https://www.aclu.org/files/dronefoia/dod/drone_dod_3160_01.pdf.

27. Chris Woods, *Sudden Justice, America's Secret Drone Wars* (New York: Oxford University Press, 2013), 220.

28. Bill Yenne, *Birds of Prey: Predators, Reapers, and America's Newest UAVs in Combat* (North Branch, MN: Specialty Books, 2010), 5.

29. Chris Woods, *Sudden Justice, America's Secret Drone Wars* (New York: Oxford University Press, 2013), 220.

30. Chris Woods, *Sudden Justice, America's Secret Drone Wars* (New York: Oxford University Press, 2013), 221.

31. I have repeatedly asked Air Force personnel and administration for this data and have been denied.

32. Alice K. Ross, "Who is dying in Afghanistan's 1000-plus drone strikes?" *The Bureau of Investigative Journalism*, July 24, 2014, accessed August 4, 2015. https://www.thebureauinvestigates.com/2014/07/24/who-is-dying-in-afghanistans-1000-plus-drone-strikes/.

33. *United Nations Assistance Mission in Afghanistan (UNAMA),* "Afghanistan Midyear Report 2014, Protection of Civilians in Armed Conflict," July 14, 2015, accessed August 3, 2015. http://unama.unmissions.org/LinkClick.aspx?fileticket=m_XyrUQDKZg%3d&tabid=12254&mid=15756&language=en-US.

34. Chris Woods, *Sudden Justice, America's Secret Drone Wars* (New York: Oxford University Press, 2013), 250–251.

35. Jon Boone, "US troops' killing of Hamid Karzai's cousin brings claim of 'deep conspiracy'," *The Guardian*, March 10, 2011, accessed August 4, 2015, http://www.theguardian.com/world/2011/mar/10/hamid-karzai-cousin-nato-death-conspiracy-afghanistan.

36. David Zucchino, "Afghans describe relatives' deaths in recent U.S. drone strike," *Los Angeles Times*, December 1, 2013, accessed August 4, 2015, http://articles.latimes.com/2013/dec/01/world/la-fg-afghanistan-drones-20131201.

37. Center for Civilians in Conflict, "Caring for Their Own: A Stronger Afghani Response to Civilian Harm," 2013, accessed August 13, 2015, http://civiliansinconflict.org/uploads/files/publications/Afghan_Report_2013_smaller_final.pdf.

38. David Rohde, "My Guards Absolutely Feared Drones," in *Drone Wars*, eds. Peter L. Bergen and Daniel Rothenberg (Cambridge: Cambridge University Press, 2015), 10.

39. The Asia Foundation, "Afghanistan in 2014: A Survey of the Afghan People," November 18, 2014, accessed November 2, 2015, http://asiafoundation.org/resources/pdfs/Afghanistanin2014final.pdf.

40. The Asia Foundation, "Afghanistan in 2014: A Survey of the Afghan People," November 18, 2014, accessed November 2, 2015, http://asiafoundation.org/resources/pdfs/Afghanistanin2014final.pdf.

41. The Asia Foundation, "Afghanistan in 2014: A Survey of the Afghan People," November 18, 2014, accessed November 2, 2015, http://asiafoundation.org/resources/pdfs/Afghanistanin2014final.pdf.

42. David Rohde, "My Guards Absolutely Feared Drones," in *Drone Wars*, eds. Peter L. Bergen and Daniel Rothenberg (Cambridge: Cambridge University Press, 2015), 10.

43. David Rohde, "My Guards Absolutely Feared Drones," in *Drone Wars*, eds. Peter L. Bergen and Daniel Rothenberg (Cambridge: Cambridge University Press, 2015), 10–11.

44. Aimal Faizi, "Drones kill any chance of peace in Afghanistan," Aljazeera, June 29, 2015, accessed August 13, 2015, http://www.aljazeera.com/indepth/opinion/2015/06/drones-kill-chance-peace-afghanistan-150628132016117.html.

45. *Center for Civilians in Conflict*, "Caring for Their Own: A Stronger Afghani Response to Civilian Harm," 2013, accessed August 13, 2015, http://civiliansinconflict.org/uploads/files/publications/Afghan_Report_2013_smaller_final.pdf.

46. *Center for Civilians in Conflict*, "Caring for Their Own: A Stronger Afghani Response to Civilian Harm," 2013, accessed August 13, 2015, http://civiliansinconflict.org/uploads/files/publications/Afghan_Report_2013_smaller_final.pdf.

47. Alissa J. Rubin, "U.S. Report on Afghan War Finds Few Gains in 6 Months," *The New York Times*, April 29, 2010, Accessed September 16, 2015, http://www.nytimes.com/2010/04/30/world/asia/30afghan.html?_r=0.

48. Anand Gopal, "The Taliban in Kandahar," in *Talibanistan*, eds. Peter Bergen and Katherine Tiedemann (New York: Oxford University Press, 2013), 6.

49. Fernando Gentilini, *Afghan Lessons, Culture, Diplomacy, and Counterinsurgency*, trans. Angela Arnone (Washington D.C.: Brookings Institution Press, 2013), 142.

50. Anand Gopal, "The Taliban in Kandahar," in *Talibanistan*, eds. Peter Bergen and Katherine Tiedemann (New York: Oxford University Press, 2013), 56–60.

51. Martine van Bijlert, "The Taliban in Zabul and Uruzgan," in *Talibanistan*, eds. Peter Bergen and Katherine Tiedemann (New York: Oxford University Press, 2013), 94–127, 116.

52. David A. Jaeger and Zahra Siddique, "Are Drone Strikes Effective in Afghanistan and Pakistan? On the Dynamics of Violence between the United States and the Taliban," *Institute for the Study of Labor:* accessed September 25, 2014, http://ftp.iza.org/dp6262.pdf, 2.

53. David A. Jaeger and Zahra Siddique, "Are Drone Strikes Effective in Afghanistan and Pakistan? On the Dynamics of Violence between the United States and the Taliban," 12.

54. American Drone strikes for 2013 and 2014 in Afghanistan are not available because they are still classified by the United States Air Force.

55. David A. Jaeger and Zahra Siddique, "Are Drone Strikes Effective in Afghanistan and Pakistan? On the Dynamics of Violence between the United States and the Taliban," 12.

56. Islamic Emirate of Afghanistan, "Statement of Islamic Emirate regarding prolonging stay of American invaders in Afghanistan," October 16, 2015, accessed November 11, 2015, http://shahamat-english.com/statement-of-islamic-emirate-regarding-prolonging-stay-of-american-invaders-in-afghanistan/.

57. Lawrence Lewis, "Drone Strikes: Civilian Casualty Considerations Executive Summary," *Joint and Coalition Operational Analysis*, accessed September 16, 2014. http://cna.org/sites/default/files/research/Drone_Strikes.pdf (This source has since been removed from the Internet).

58. Chris Woods, *Sudden Justice, America's Secret Drone Wars* (New York: Oxford University Press, 2013), 224.

59. Anthony Shaffer, *Operation Dark Heart: Spycraft and Special Ops on the Frontlines of Afghanistan and the Path to Victory* (New York: Thomas Dunne Books, 2010), 284.

60. Anthony Shaffer, *Operation Dark Heart: Spycraft and Special Ops on the Frontlines of Afghanistan and the Path to Victory* (New York: Thomas Dunne Books, 2010).

Chapter 3

Pakistan

America's Best Ally and Worst Enemy

The introductory Chapters 1 and 2 on Afghanistan included histories of the Taliban and al-Qa'ida in Afghanistan, thus this information will not be repeated in this chapter. The border relationship between Afghanistan and Pakistan will be examined as it is imperative to understanding how both al-Qa'ida and the Taliban operate. Furthermore, the relationship between the Taliban in Afghanistan and the Taliban in Pakistan will be discussed briefly as it is pertinent to understanding American targeted killing in Pakistan. Lastly, the relationship between the Haqqani Network and the Pakistani Government will be addressed. This chapter will then examine the results of American targeted killing in Pakistan.

The histories of Pakistan and Afghanistan are closely intertwined. The Durand Line which led to the separation of Afghanistan and Pakistan was drawn in 1893 by the British Government of Colonial India and Abdur Rahman Khan, Emir of Afghanistan on November 12, 1893. The line almost divides the Pashtun tribal area into two equal parts and is approximately 2430 km across the middle of the Pashtun territory. The Pashtun area predominately houses Taliban and al-Qa'ida sites and is the main American drone strike area. It comprises approximately thirty-five million Pashtun in Pakistan, which is somewhere around 20 percent of Pakistan's total population. In Afghanistan, the Pashtun make up about 50 percent of the total population or twenty million people.[1]

Modern Pashtun nationalism emerged in 1930 when British troops killed unarmed troops in the city of Peshwar. A nationalist movement lead by Khan Abdul Ghaffar Khan eventually found its way into Congress, hoping to create an independent Pashtun country. In August 1947, India and Pakistan became two new countries after they left the British Indian Empire. Shortly afterwards, the Afghanistan Government refused to vote for Pakistan's admission to the United Nations in 1947 but withdrew its challenge only after one month. Afghanistan subsequently became the first country to recognize Pakistan. In 1947, Pakistan began fighting India for Kashmir, a Muslim state belonging to India, after developing an independent military. Pakistan has fought over Kashmir ever since.

Afghanistan has used the Pashtun cause to try and rally support for the Afghanistan state. This support of the Pashtuns led to ethnic rivalry between the Pashtuns and the non-Pashtuns in Afghanistan and has caused Pakistani resentment. Al-Qa'ida and the Taliban have also jumped at the chance to exacerbate these social cleavages. In 1961, Pakistan closed its border and Afghanistan and Pakistan broke off diplomatic relations. A tit for tat cycle has ensued throughout the history of diplomatic relations between Pakistan and Afghanistan, although Pakistan has always been the stronger state. Pakistan has used the constant conflict in Afghanistan for its own benefit and has relished the benefits that instability and lawlessness have provided.[2] Both countries have continued to provide economic support and sanctuary to dissidents and terrorists so frequently that it has led to a diplomatic cessation on several occasions.

Taliban and Pashtun people cross the border regularly between Pakistan and Afghanistan without much attention paid to border security or control. This lack of security and monitoring of border crossings has made it extremely difficult to track al-Qa'ida, al-Haqqani, and Taliban members. In fact, many of these members have lived in caves. in this area and have literally disappeared in the tumultuous region.

The Pakistani Government has played a role in the funding and direction of the Taliban. Initially they funded the Taliban to destabilize the Soviets and to deploy troops if necessary over the Kashmir dispute with India. However, as time progressed, several sects within the government have developed varying expectations for the Taliban. Many Pakistanis and military personnel see the Taliban as freedom fighters fighting the invading Americans, just like the Russians in Afghanistan during the Cold War. Currently, Pakistan is a safe haven for Taliban who are fleeing American and NATO forces. Taliban are able to cross over the border into Pakistan from Afghanistan and find protection and support.[3] Pakistani military members have stated, "It's not that we like the Taliben, but they are all we've got."[4] The Taliban has managed to establish a foundation in Pakistan and managed to institute Sharia Law in the FATA area of Pakistan although the Taliban in Pakistan exist as a territory of the Taliban in Afghanistan.[5]

From the perspective of the Pakistani Government, September 11, 2001 resulted in tremendous changes for Pakistan. Former Pakistani President Musharraf talked about 9–11 in his memoir *In the Line of Fire*. Immediately after the event, Musharraf was concerned as Pakistan was the only country that had diplomatic relations with the Taliban. On September 13, 2001, Wendy Chamberlain, the Ambassador to Pakistan, brought Musharraf a list of seven demands from the United States' President. These demands included stopping al-Qa'ida operatives, an open fly zone over Pakistan, territorial access to Pakistan, Pakistani Government's public condemnation of 9–11, the rescindment of fuel shipments to the Taliban, and a promise to share intelligence with the Americans.[6] Although Musharraf stated that he was frustrated and dismayed by these demands, he endeavored to appease the American Government and the Bush Administration.

After September, drone activity in Pakistan began to pick up. In 2002 to June 2004, Predators simply patrolled the Pakistani skies, searching for information. Then on June 18, 2004, Nek Muhammad, the Pakistani Taliban leader was traced and killed while talking on his cell phone in South Waziristan.[7] The next strike came on May 14, 2005 and killed Haitham al-Yemini, a high-ranking al Qa'ida weapons expert in North Waziristan.[8] On December 3, 2005, the number three leader in al-Qa'ida was killed, Egyptian-born Abu Hamza Rabia.[9] The drone strikes from then on continued to trickle on until late 2007 and January, 2008, when drone strikes in Pakistan became a several times a month occurrence. This time period correlates with the takeover of drone strikes in Pakistan by the US Army's Joint Special Operations Command (JSOC).[10]

However, the relationship between the Pakistani Government and Haqqani, or even the Taliban, is not that simple to understand. While in public, the Pakistanis condemn the Taliban and the Haqqani, but covert support is evident. The Pakistani Government shares a symbiotic relationship with several combative terrorist organizations. The Haqqani Network in particular, is used by the ISI (Pakistan Intelligence Network) to provide covert information and as a state asset at opportune moments. Pakistani senior state officials offered the Haqqani Network as a bargaining chip to reconcile relations with Pakistan when the United States was preparing to leave Afghanistan. The 2009 former ISI Chief Hamid Gul considered the leader of the Haqqani Network, Jalaluddin Haqqani "a personal friend" and stated that he made sure that his two sons who were sent to Afghanistan to fight the Soviets fought alongside "Haqqani's" men.[11] ISI also uses the Haqqanis to counter Indian activity and influence in Afghanistan, mostly in the capital Kabul and to create conflict with the United States and NATO in Afghanistan. In return, Pakistan protects the Haqqani Network including providing them with warning of raids and drone attacks. ISI also provides Haqqanni with logistics, materials, advising, and training.[12] A similar relationship exists with the Taliban in Afghanistan and Pakistan. The next section provides the results of American targeted killing in Pakistan.

THE RESULTS OF AMERICAN TARGETED KILLING IN PAKISTAN

The first component in addressing the effects of American drone strikes in Pakistan is the issue of sovereignty. Technically the Pakistani government has not publicly given its consent for American drone strikes in Pakistan, but this strange arrangement belies an underlying agreement between the Pakistani Government and the American Government. In April 2013, President Pervez Musharraf stated that he gave consent for the United States to conduct drone strikes only two or three times. Musharraf stated that "Only on a few occasions, when a target was absolutely isolated and [there was] no chance of collateral damage" did he give the Americans consent for a drone strike.[13] However the Bush Administration then denied incompliance and US Media Firm McClatchy stated that "That partnership was so extensive during the Bush administration that the Pakistani intelligence agency selected its own targets for drone strikes. Until mid-2008, the CIA had to obtain advance approval before each attack, and under both [the Bush and Obama Administrations] the Pakistanis received briefings and videos of the strikes."[14]

Under the Obama Administration, ISI is unable to veto any decision on drone strikes. Although President Obama states that he has Pakistani compliance to conduct drone strikes at will in Pakistan, the Pakistani Congress vehemently denies this. President Asif Ali Zardari, who was president of Pakistan from 2008 to 2013, stated that drone strikes were "counter-productive and violated Pakistan's sovereignty."[15] President Mahmood Hussain, elected on September 9, 2013 has a different view of drone strikes than President Musharraf. "[Drone strikes] are counterproductive, [they] fuel anti-American sentiment in Pakistan, and breed more terrorism."[16] President Hussain publicly condemns the drone strikes but is not at liberty to do anything about it either as drone strikes come with significant amounts of American aid. Pakistan usually receives around one billion per year in American aid. Congress threatened to withhold 300 million in 2015 if Pakistan did not do more to disrupt the Haqqani Network.[17]

The Obama Administration states that drone strikes are used to kill high-ranking members of al-Qa'ida and other affiliated groups. In Pakistan, the high-profile strikes so far have killed the founder of the Pakistan Taliban, Baitullah Mehsud and his successor Hakimullah Mehsud. The al-Qa'ida second-in-command Abu Yahya al Libi was also killed. President Obama's Administration has been responsible for the deaths of 2379 people in Pakistani as of October 21, 2015.

The Pakistani military's attitude toward the drone strikes is somewhat volatile. The military leadership has sought control over drone strikes, which discontinued under Obama, to punish enemies. Sometimes, the military has tried to protect members of the Haqqani Network from drone strikes as these people were allies. In November 2011, relations ended with the NATO attack that killed twenty-four Pakistani soldiers and the government demanded that the United States leaves a military base, which it left Shamsi in December 2011.[18]

The effects of drone targeted killing on Pakistani civilian population are not encouraging but instead provide a litany of problems such as: a long list of victims (collateral damage), Pakistani support for terrorist groups, an increase in anti-Americanism, a dispersal of terrorist group elements throughout Pakistan, psychological consequences for Pakistanis, and an increase in suicide terrorist attacks among many other problems.

In Figure 3.1: *American Drone Strikes in Pakistan, 2004–2015*,[19] the results of American drone strikes in Pakistan are shown. Between 2004 and 2015, approximately 429 American strikes occurred and over 2500 people were killed according to *The Bureau of Investigative Journalism*, including the targets. These strikes resulted in the deaths of almost six people per drone strike, again including the targets. It is highly unlikely that only six people were targeted per drone strike, implying a lot of collateral damage.

The "Naming the Dead Project" created by *The Bureau of Investigative Journalism* provides a list of people reportedly killed by American drone strikes in Pakistan. "Over the past decade, more than 2500 people have been killed by the CIA drone strikes in the tribal areas of north west Pakistan. . . . About a fifth of the dead are described as civilians."[20] Of those 2500 people,

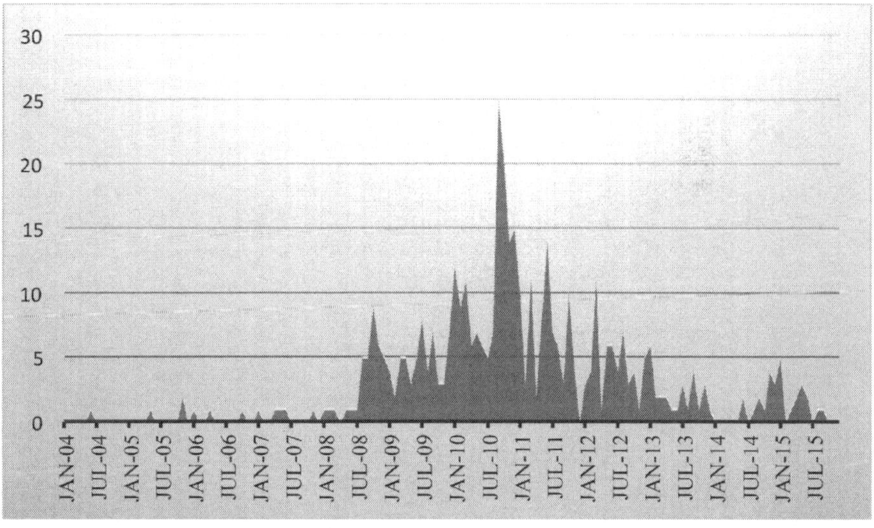

Figure 3.1 American Drone Strikes in Pakistan, 2004–2015. *Source*: *The Bureau of Investigative Journalism*, "Naming the Dead," accessed September 16, 2014, http://www.thebureauinvestigates.com/namingthedead/the-dead/?lang=en.

ninety nine were children. Each person's case is detailed and chronicled in the "Naming the Dead Project." Most likely, it can be assumed that Pakistani children are collateral damage from drones not members of terrorists groups.

Pakistan is home to several terrorist organizations and is one of the most essential countries in the War on Terror, pertinent to the defeat of al-Qa'ida because of its proximity to Afghanistan and its history as an al-Qa'ida base of recruitment and financing. Al-Qa'ida inspires other groups such as: Quetta Shura, the Haqqani Network, the Pakistan Taliban, and LeT & Co among other groups[21] that have also used terrorism as a means of political change. Drone strikes have provided an increase in recruits and empathy for these terrorist groups. Unfortunately, numbers of terrorist groups can only be estimated not researched as this information does not exist, so anecdotal evidence must be used.

In an early article on drones in 2013, Michael Boyle argues that drone strikes "corrode the stability and legitimacy of local governments, deepen anti-American sentiment and create new recruits for Islamist networks aiming to overthrow these governments."[22] Imran Khan, political opponent of the Pakistan People's Party and the chief of Pakistan's Tehrik-i-Insaf Party has stated that he would shoot American drones down if he became president. Khan insists "that the drone missions only serve to further radicalize young Pakistanis." "[Drones are] totally counterproductive," he told BBC. "The [idea] that it is only killing al Qaeda is a myth. The people will tell you that the vast majority of people killed are either innocent civilians or some sort of low-level militants."[23] Tehreek-e-Insaf (PTI) is devoted to the discontinuation of American drone warfare and targeted killing so much so that it is their political platform.[24,25]

The rival part of the PTI, the Pakistan People's Party (PPP) has similar anti-American views. A PPP member in the National Assembly stated in April 2009:

> We used to be very liberal, pro-Western people, but American behavior and attitudes are forcing us to develop our own identity, because we cannot simply be your servants. The Taliban are religious fanatics but so is Bush and many Americans. Worst of all, the Americans are forcing us to make mistakes and we are suffering as a result, and yet still they are blaming us for not doing enough. America faced only one 9/11. Due to our helping America, we in Pakistan are now facing 9/11s continuously with so many dead, and American policies are continuously making things worse, killing people, helping the Taleban and spreading disorder. . . . Many people here think the reason can only be that the Americans are creating all this disorder deliberately because they want to establish military bases here against China and Iran.[26]

Wajid Shamsul Hasan, the high commissioner to London and one of Pakistan's top ambassadors states that American drone strikes weaken

Pakistan's democratic institutions and increase terrorist empathy. Hasan has vowed to stop the drone strikes at any cost.

> Even those who were supporting us in the border areas have now become our enemies. They say that we are partners in these crimes against the people. By and large you will hardly find anybody who will say a word in support of the United States, because of these drone attacks.[27]

THE EFFECT OF DRONES ON THE PAKISTAN PEOPLE

Pew public opinion polls find that support for al Qa'ida, the Taliban, Tehrik-i-Taliban, the Afghan Taliban, Lashkare-Taiba, and the Haqqani have not really changed over the last few years. As can be seen in Table 3.1: *Pakistani Support for Terrorist Groups, 2008–2012,* there is little variance in recent years. Approximately 10 to 25 percent of Pakistanis, depending on the year and terrorist group, support terrorist groups. Conversely almost a third of Pakistanis interviewed do not know how they view the terrorist groups or are possibly refusing to say anything out of fear.

In addition to anti-drone Pakistani political parties and ambassadors, drone strikes in Pakistan have caused many anti-U.S. street demonstrations and an increase in anti-American sentiment.[28] Drone strikes in one village in January

Table 3.1 Pakistani Support for Terrorist Groups, 2008–2012

Al Qaeda	*Favorable*	*Unfavorable*	*Don't Know*	*The Taliban*	*Favorable*	*Unfavorable*	*Don't know*
2012	13	55	31	2012	13	66	20
2011	12	55	33	2011	12	63	24
2010	18	53	28	2010	15	65	29
2009	9	61	30	2009	10	70	20
2008	25	34	41	2008	27	33	40
Tehrik-i-Taliban	*Favorable*	*Unfavorable*	*Don't Know*	*Afghan Taliban*	*Favorable*	*Unfavorable*	*Don't know*
2012	17	52	32	2012	14	45	41
2011	19	51	30	2011	15	50	35
2010	18	51	31	2010	16	49	34
Lashkar-e-Taiba	*Favorable*	*Unfavorable*	*Don't Know*	*Haqqani Network*	*Favorable*	*Unfavorable*	*Don't Know*
2012	22	37	41	2012	5	31	64
2011	27	37	36				
2010	25	35	40				

Source: Pew Research Center, "Pakistani Public Opinion Ever More Critical of U.S." June 27, 2012, accessed January 27, 2015, http://www.pewglobal.org/2012/06/27/pakistani-public-opinion-ever-more-critical-of-u-s/.

2006 led to the deaths of eighteen people in Damadola, including five women and five children. None of the people killed belonged to al-Qa'ida. The drones, according to locals that had been seen flying over the village for the previous three days were aimed at Ayman al-Zawahiri. In response to the drone strikes, on January 12, 2006 over 8000 tribesmen, led by Haroon Rasheed, a member of the radical Islamic Party *Jamate Islami* protested the drone strikes that killed eighteen innocent people.[29] Pakistan's civilian government is weak and walks a fine line between its alliance with the United States' Government and protecting its citizens. Pakistani expert Ahmed Rashid, states that "The drones have become the target of widespread public anger at the government, the army and the Americans."[30]

An early public opinion poll in the Spring 2009, from the Pakistan-based Aryana Institute for Regional Research and Advocacy found that:

- 45 percent of the Pashtuns questioned felt that the drone strikes brought fear and terror to common people.
- 52 percent questioned felt the strikes were accurate.
- 58 said that the strikes did not cause anti-Americanism.
- 60 percent felt that the militants were damaged by the strikes.
- 70 percent felt that the Pakistanis should carry out strikes of their own against the militants.[31]

In Table 3.2: *Pew Polling in Pakistan 2011,* 62 percent of Pakistanis oppose "U.S.-led efforts to fight terrorism" while only 16 percent of the population supports the efforts. Fifty six percent of Pakistanis think that the United States does not consider Pakistani interests while only 20 percent think that the Americans do consider Pakistani interests. According to the Pew Research Global Attitudes Project, 3 percent of Pakistanis "approve" and 66 percent "disapprove." of the U.S. using drone strikes to target extremists."[32] One humanitarian worker in Waziristan stated,

> Do you remember 9–11? Do you remember what it felt like right after? I was in New York on 9/11. I remember people crying in the streets. People were afraid about what might happen next. People didn't know if there would be another attack. There was tension in the air. This is what it [living with the constant threat of drone attacks] is like. It is a continuous tension, a feeling of continuous uneasiness. We are scared. You wake up with a start to every noise.[33]

A study conducted by Fair, Kaltenthaler, and Miller[34] of 7,656 Pakistanis in 2013, found that 50 percent stated that they did not support American drone strikes aimed at militants in the FATA. Three percent supported it a lot and 7 percent somewhat supported it. Thirty-five percent did not know. Although the purpose of this study was to understand how Pakistanis get data, the

Table 3.2 Pew Polling in Pakistan 2011

U.S. Considers interests of countries like Pakistan?	
Great Deal/Fair Amount	20%
Not too much/Not at all	56%
Don't Know	24%
U.S. and NATO Troops in Afghanistan	
Keep Troops	8%
Remove Troops	69%
Don't Know	22%
U.S.-led Efforts to Fight Terrorism	
Favor	16%
Oppose	62%
Don't Know	22%

Source: Pew Research Global Attitudes Project, "U.S. Image in Pakistan Falls No Further Following bin Laden Killing Support for Campaign against Extremists Wanes," June 21, 2011, accessed October 12, 2014, http://www.pewglobal.org/2011/06/21/u-s-image-in-pakistan-falls-no-further-following-bin-laden-killing/.

results imply that Pakistanis do not like American drone strikes, even though Pakistanis primarily get information by word of mouth.

These numbers have continued to decline since 2014 according to Pew polling. Around 66.6 percent of Pakistanis oppose U.S. drone strikes. Only 3 percent approve and 30 percent do not have an opinion. Another 66.6 percent of Pakistanis think that drone strikes kill too many innocent people and only 21 percent think they are necessary to defend Pakistan. Forty-one percent of Pakistanis believe that drone strikes are being conducted without the consent of the Pakistani Government.[35]

In addition, American drones have caused widespread psychological paranoia and hysteria. Pakistan locals call drones "*bangana*" because of the constant noise that they make when they fly above.[36] This constant buzzing has had a severe psychological toll on Pakistanis. There are reports of children who become hysterical when they hear drones buzzing above their heads. The *Center for Civilians in Conflict* was told by one victim, "We fear that the drones will strike us again . . . my aged parents are often in a state of fear. We are depressed, anxious, and constantly remembering our deceased family members . . . it often compels me to leave this place." Another man described his sister-in-law, who lost her husbands and two sons in American drone strikes. "After their death she is mentally upset . . . she is always screaming and shouting at night and demanding me to take her to their graves." Some civilians who used to see one drone per week under the Bush Administration now report seeing as many as ten to fifteen per day.[37] Although more research needs to be completed on this, there appears to be some widely spread post-traumatic stress among the Pakistanis as a result of drone targeted killing.

THE DISPERSAL OF TERRORIST GROUP ELEMENTS

Joshua Foust states that a successful drone strike changes the behavior of terrorist organizations in three possible ways. Theses behavioral modifications include: "rejecting technology, going into hiding, and violently attacking those suspected of participating in the targeting process."[38] It is true that American drone strikes in Pakistan have caused terrorist groups such as al-Qa'ida and the Taliban to reject technology and to go into hiding, including moving to more populous cities and other countries. Phones and computers are used less frequently as they are inherent in tracking devices. Drone attacks not only cause terrorist elements to go into hiding but they also cause dispersal to other areas.

In an interview on September 13, 2007, directly after the American drone wars began in Pakistan, Sociologist Rosemarie Skaine asked Professor Abdul Raheem Yaseer, the Assistant Director for the Center for Afghanistan Studies at the University of Nebraska at Omaha about the movements of the Taliban in Afghanistan. Professor Yaseer stated,

> As you know, the Taliban were supposed to be totally eliminated. They were not eliminated. They were just pushed out of Afghanistan into Pakistan where they found safe havens with the support of the Pakistani military. They had camps and schools and madrasas where they got more training. Their training did not stop, it continued. They received more money and arms and facilities. Therefore, they were not totally eliminated. As a matter of fact, they were supported. Then when the attention was diverted from Afghanistan to Iraq that provided a good opportunity for the Taliban who went into Pakistan to get reunited. Their supporters Saudi Arabia being part of [this group and the] Pakistani military [being another] part of it, and some other elements here and there, helped them to get reorganized, to regroup and then they were allowed to infilitrate into Afghanistan and kill teachers, warn women and burn schools. They caused insecurity and issued threats.[39]

Professor Wali Aslam published a report for *The Remote Control Project* concerning the unintended consequences of drone strikes in Pakistan. Al-Qa'ida and Taliban are leaving rural FATA areas in droves and are flocking to major cities including Karachi to evade drone strikes, creating problems such as communal and sectarian violence, gang warfare, an increase in extremist Madrassas, kidnapping for ransom, land grabbing, political assassination, and fierce competition for already scarce resources.[40] The relocation of terrorist elements in Pakistan are not part of any logical counterterrorism strategy but instead lead to the dispersal of terrorists throughout Pakistan. Drone strikes are causing the terrorist to disperse and create unrest in other areas, spreading violence and crime.

AN INCREASE IN SUICIDE BOMBINGS IN PAKISTAN

Most importantly, American drone strikes are highly correlated with suicide bombing attacks in Pakistan. In Figure 3.2: *Drone Strikes and Suicide Bombings in Pakistan January 2004–September 2015,* it is evident that there is a correlation between drone strikes and suicide bombings.

To take this a step further using simple linear regression (OLS) as can be seen in Figure 3.1: *American Drone Attacks and Suicide Bombing January 2004–September 2015*, there is a positive relationship between suicide bombings in Pakistan and American drone strikes in Pakistan per month. For every suicide bombing, there is a 14.9 increase in American drone strikes per month, which makes sense. For every American drone strike, there is a 41.9 increase in suicide bombings per month. This is counterintuitive, especially if we believe that drone strikes are supposed to stop suicide terrorist attacks throughout the world. Instead, an American drone strike is more likely to cause a suicide bombing. This may be explained by the fact that terrorists become vengeful after seeing their property, home, friends, or family destroyed. Although this regression is bivariate, it is important to include in the chapter because it belies a simple strong relationship between two variables, which is demonstrated first in Table 3.3: *American Drone Strikes and Suicide Bombings in Pakistan January 2004–September 2015.*[41]

Besides the civilian death toll, several other unintended consequences have appeared as a result of the drone strikes in Pakistan. The United States has experienced suicide bombing attacks in retaliation for drone strikes in

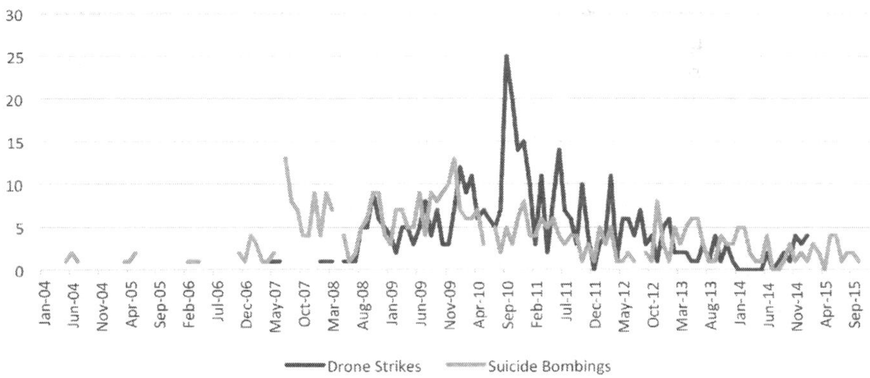

Figure 3.2 Drone Strikes and Suicide Bombings in Pakistan: January 2004–September 2015. *Source*: For Pakistan suicide bombings: Pakistan Body Count.org, "Pakistan Body Count," 2015. accessed September 16, 2015, http://pakistanbodycount.org/suicide_bombing; For U.S. drone attacks on Pakistan: The Center for the Study of Targeted Killing, accessed September 16, 2014, http://www.targetedkilling.org/.

Table 3.3 American Drone Attacks and Suicide Bombings January 2004–September 2015

Independent Variable	*Dependent Variable*
American Drone Attacks	Suicide Bombings .149** (.062)
Suicide Bombings	American Drone Attacks .419** (.174)
Observations	89

Standard errors in parentheses and $**p < 0.05$
Source: For Pakistan suicide bombings: Pakistan Body Count.org, Pakistan Body Count, accessed September 16, 2015, http://pakistanbodycount.org/suicide_bombing.

Pakistan. The Khost Bombing in December 2009 was committed by Humam Khalil Abu Mulal al-Balawi, a Jordanian who walked into a CIA compound at Camp Chapman, located outside Khost, Afghanistan. Al-Balawi killed seven intelligence officers and one Jordanian intelligence officer. Although al-Balawi was recruited as a spy, in the video released after his death, al-Balawi stated, "This attack will be the first of revenge operations against Americans and their drone teams outside Pakistani borders." Al-Balawi was avenging the death of Beitullah Mehsud who was killed by a drone strike in Pakistan in August 2009.[42] On January 14, 2015, the US Airforce launched a drone strike against an al-Qa'ida compound near the Afghanistan border and inadvertently killed Warren Weinstein, an American professor and citizen, and an Italian citizen, Giovanni lo Porto.[43]

American targeted killing has also led to local purification squads in areas where drone attacks occur. Al-Qa'ida or Taliban members will patrol an area where a drone strike or surveillance has occurred and will question the population. The CIA frequently offers monetary rewards to people who inform on al-Qa'ida and Taliban activities from 300 to 1000 dollars. Taliban and al-Qa'ida will frequently torture or kill those people who they suspect of disloyalty. Favorite forms of torture includes inserting a rod through the stomach or arms, or being suspended over a fire. These groups are called *Khorasan Mujahedin* by local populations in the Waziristan region of Pakistan.[44] Bodies will often be left beside the road with a sign attached to them stating the person was a spy. In summer, 2012, al-Qa'ida crucified a man on a lamppost who was suspected of helping the American drone campaign.[45] Most of the people who are accused of being informants are not informants but will typically confess to any accusation after being physically tortured. Al Qa'ida has become so paranoid of spies working for the Americans within their own organization that they published an online manual called *Al-Qa'ida's Guidebook on How to Catch Muslim Spies* for jihadist groups.[46]

INFAMOUS TARGETS

Several high- middle- and low-profile terrorists have been killed in Pakistan by American drones or airstrikes. Although this list may not be entirely complete, these men's deaths were published by media outlets, which is why they are listed below. They are listed by year, including occupation.

In 2015:

- Khawray Mehsud: Senior military commander, al-Qa'ida, bodyguard Baitullah Mehsud
- Ustad Ahmad Farooq: Deputy Emir for al-Qa'ida in Indian subcontinent
- Qari 'Imran: Member of al-Qa'ida executive council in Indian subcontinent
- Adam Gadahn (American citizen): Propagandist for al-Qa'ida coordinator for Arab region and Pakistan

In 2014:

- Omar Farroq-al-Qa'ida: Coordinator for Arab region and Pakistan
- Adil Abdul Quoos: Senior leader in Indian subcontinent
- Dr. Sarbaland: Surgeon and media emir
- Abdullah Haqqani: Director of suicide operations Haqqani Network
- Sheikh Imran Ali Siddiqi: Senior leader in Indian subcontinent
- Muhammad Mustafa: Senior commander in Hafiz Gul Bahadar Group
- Taj al Makki: Mid-level commander
- Abu Abdurahman al Kuwaiti: Mid-level commander
- Fayez Awda al Khalidi: Mid-level commander
- Haji Gul-Haqqani: Network commander
- Mufti Sofian-Haqqani: Network commander
- Commander Abu Bakr: Senior commander in Afghanistan Taliban

In 2013:

- Abdul Rehman: Senior Haqqani Network commander
- Mufti Hamidullah Haqqani: Senior Haqqani Network commander
- Maulvi Ahmed Jan: Senior Haqqani Network commander
- Hakeemullah Mehdsud: Emir Taliban in Pakistan
- Mullah Sangeen Zadran: Deputy in Haqqani Network
- Abu Rashid: Mid-level al-Qa'ida military trainers
- Muhammed Ilyas Kuwaiti: Mid-level al-Qa'ida military trainers
- Muhammed Sajid Yamani: Mid-level al-Qa'ida military trainers
- Abu Saif al Jaziri: Senior al-Qa'ida military trainer
- Maulana Akhtar Zadran: Haqqani Network commander

- Waliur Rehman: Deputy Emir Taliban in Pakistan
- Abu Ubaydah Abdullah al Adam: Al-Qa'ida chief propagandist for sect Vanguards of the Khorasan
- Sheikh Yasin Al Kuwaiti: Paramilitary commander in Shadow Army in al-Qa'ida
- Wali Mohammed: Said to have operated Pakistan Taliban operations
- Faisal Khan: Mid-level commander in Hakeemullah's Mehsud's Movement (Taliban)
- Mullah Nazir: Led local Taliban faction (was a Pakistani ally)[47]
- Atta Ullah: Deputy of Nazir
- Rafey Khan: Deputy of Nazir
- Rata Khan: Senior military commander

In 2012:

- Mohammad Ahmed al Mansoor: Mid-level al-Qa'ida commander
- Khlaid bin Abdul Rahman al Husainan: Cleric
- Abdul Rehman al Zaman Yemeni: Mid-level al-Qa'ida commander
- Sheikh Abdul Bari: Mid-level al-Qa'ida commander
- Hasan Ghul: Osama bin Laden's emissary to Abu Musab al Zarqawi
- Abum Kasha al Iraqi: Al-Qa'ida commander in Mir Ali area
- Fateh al Turki: Senior Al-Qa'ida military leader
- Badruddin Haqqani: Deputy and brother of Sirajuddin Haqqani
- Abdul Shakoor Turkistani: Leader of Turkistan Islamic Party and al-Qa'ida operations chief
- Angineer Ahsan Aziz: Kashmiri jihadist who served as al-Qa'ida commander
- Abu Yahya al Libi: Al-Qa'ida's chief of staff and cleric
- Abu Usman Adil: Leader of Islamic movement of Uzbekistan
- Badr Mansoor: Al-Qa'ida leader linking Taliban and Pakistani jihadist groups
- Aslam Awan: Deputy to al-Qa'ida leader of external operations
- Qari Hussain Mehsud: Leader in Pakistan Taliban

In 2011:

- Hazrat Oamr: Mullah Nazir's brother and commander
- Khan Mohammed: Senior deputy to Nazir
- Miraj Wazir: Senior commander Taliban
- Ashfaq Wazir: Senior commander Taliban
- Abu Miqdad al Masri-al-Qa'ida Shura Majlis member
- Abd al Rahman al Yemeni: A senior operative in al-Qa'ida's external operations network
- Jan Baz Zadran: Siraj haqqani's deputy
- Haleem Ullah: Deputy commander to Taliban leader Hafiz Gul Bahadar

- Abu Hafs al-Shahri: al-Qa'ida operations chief Pakistan
- Atiyah Abd al-Rahman: Osama bin Laden's chief of staff
- Ilyas Kashmiri: Leader of al-Qa'ida's Lashkar al-Zil
- Abu Zaid al Iraqi: Financial advisor to al-Qa'ida

In 2010:

- Ibn Amin: Senior commander of forces
- Mohammed Usman: Member of Iylas Kashmiri's Brigade 313, key aide to Osama bin Laden
- Sheikh Fateh al Masri: al-Qa'ida's leader in Afghanistan and Pakistan
- Saifullah Haqqani: Haqqani Network commander
- Qureshi: Trainor of jihadists
- Inayatullah: Taliban military commander in North Waziristan
- Hamza al Jawfi: Leader of Jundallah
- Abu Ahmed: al-Qa'ida commander in Afghanistan
- Sheikh Ihsanullah: al-Qa'ida commander in Afghanistan
- Ibrahim: Commander of Fursan-i-Mohammed Group
- Osama bin Ali bin Abdullah bin Damjan al Dawsari: A Senior Taliban operative
- Mustafa Abu Yazid: al Qa'ida general manager
- Sadam Hussein Al Hussami: Senior al-Qa'ida operative
- Qari Mohammed Zafar: Leader of Feyadeen-i-Islam
- Mohammed Haqqani: Mid-level Haqqani Network commander
- Sheikh Mansoor: al-Qa'ida Shadow Army commander
- Jamal Saeed Abdul Rahim Mansur al Shami: Ideologue and aid to Mustafa Abu Yasid

In 2009:

- Abdullah Said al Libi: Commander of Lashkar al-Zil, al-Qa'ida's Shadow Army
- Zuhaib al Zahib: Commander Lashkar al-Zil, al-Qa'ida's Shadow Army
- Saleh al Somali: Leader of al-Qa'ida's external network
- Abu Musa al Masri: Senior al-Qa'ida explosive expert and trainer
- Najmuddin Jalolov: Leader of Islamic Jihad Group
- Maulvi Ismail Khan: Military commander in Haqqani Network
- Mustafa al Jaziri: Senior military commander in al-Qa'ida, sat on shura
- Tahir Yuldashev: Leader of Islamic movement in Uzbekistan
- Baitullah Mehsud: Leader of Pakistan Taliban
- Kifayatullah Anikhel: Taliban commander
- Mufti Noor Wali: Suicide bomber trainer for Taliban and al-Qa'ida
- Khwaz Ali Mehsud: Senior deputy to Baitullah Mehsud

- Abdullah Hamas al Filistini: Senior al-Qa'ida trainer
- Osama al-Kini: al-Qa'ida's operation chief in Pakistan
- Sheikh Ahmed Salim Swedan: Senior aide to Osama al-Kini
- Sa'ad bin Laden: One of Osama bin Laden's sons who severed as a senior al-Qa'ida member
- Rashid Rauf: Senior al-Qa'ida leader[48]

In 2008:

- Abu Zubair al-Masri: al-Qa'ida explosives expert
- Abdullah Azzam al Saudi: Recruiter and trainer for al-Qa'ida
- Abu Jihad al-Masri: Leader of Egyptian Islamic
- Khalid Habib: Commander of Lashkar al-Zil or the Shadow Army
- Abu al Hasan al Rimi: Senior al-Qa'ida operative
- Abu Ubaidah al Tunisi: al-Qa'ida military commander
- Abu Masa: al-Qa'ida operative
- Abu Qasim: al-Qa'ida operative
- Abu Hamza: Explosives expert from Saudi Arabia
- Abu Haris: Senior military commander
- Abu Wafa al Saudi: al-Qa'ida commander and logistician
- Abdul Rehman: Local Taliban commander
- Abu Khabab al-Masri: chief of al-Qa'ida's WMD Program and master bomb maker
- Abu Mohammed Ibrahim bin Abi al Faraj al Masri: Religious leader
- Abdul Wahhab al Masri: A senior aide to Abu Khabab al-Masri
- Abu Islam al-Mari: Aide to Abu Khabab al-Masri
- Abu Salayman Jazairi: Chief of al-Qa'ida's external network
- Dr. Arshad Waheed: Mid-level al-Qa'ida commander
- Abu Laith al Libi: Senior military commander and leader of reformed Brigade 055 in al-Qa'ida's Shadow Army

No one was killed in 2007

In 2006:

- Liaquat Hussain: Second in command of Bajaur TNSM
- Imam Asad: Camp Commander of Black Guard (elite bodyguards for Osama bin Laden)

In 2005:

- Abu Hamza Rabia: al-Qa'ida's operational commander
- Haitham al Yemeni: A senior al-Qa'ida's explosive expert

In 2004:

- Nek Mohammed: Senior Taliban commander in South Waziristan[49]

To summarize the information presented, approximately 119 terrorists in al-Qa'ida, al-Haqqani, or the Taliban have been killed in drone strikes in Pakistan since 2004 according to all media sources. In contrast 2500 people have been killed accidently in drone strikes while targeting these terrorists. That is approximately twenty people killed per terrorist killed, not including the aforementioned terrorists. This means that around twenty people are killed for every terrorist target. Although it is definitely possible that some of the people that were killed were militants, these people were not mentioned in media reports.

Many of the terrorists that have been killed are low-level deputies or mid-level commanders. It is also suspect that several of these people that were killed occupied the same position as though American media sources did not really know what these people did. Either al-Qa'ida has replaced people very quickly or many people within the organization carry similar administrative titles. Or from another viewpoint, perhaps the American military or US media exaggerates the prestige of the terrorists they have killed. However, one does wonder what information could be gained if these suspects were captured and interrogated, instead of killed.

CONCLUSION

These findings present a dismal view of drone targeted killing in Pakistan. Pakistani public support for terrorist groups has not really decreased or increased over approximately the last five years according to public opinion polls. Drones do lead to an increase in anti-Americanism according to public opinion polling and political commentary. Most Pakistani citizens do not want any US military in the area. The most interesting finding is that American drone usage leads to an increase in suicide terrorist attacks in Pakistan while suicide attacks lead to a smaller increase in American drone attacks. Most importantly from the data, it can be said that while American drone usage does kill terrorists in the short run, it cannot bring an end to terrorist groups and terrorist attacks. Instead drone usage leads to a never-ending cycle of tit for tat between terrorist groups, governments, and the United States, thus supplying terrorist groups with a never-ending supply of recruits.

Avery Plaw and Matthew S. Fricker conclude in their article in *International Perspectives* that the drone policy in Pakistan would be much more productive if the United States more narrowly focused its campaign on al-Qa'ida and the Taliban in both Pakistan and Afghanistan. Smaller surgical strikes with greater accuracy would be more productive than what the United States and President Obama are currently doing. Large casualty rates and massive collateral damage is not creating the intended results. In addition, the author recommends, as mentioned earlier in this book, a more robust legal policy where the president stops targeting American citizens and is not the sole decision-maker concerning targets. Institutional protocol and chain of command for drone strikes would be an improvement.[50]

One of the biggest problems with fighting terrorism in Pakistan is the support that the Pakistani military and Pakistani intelligence agencies have for al-Qa'ida and the Taliban. In his book, *Operation Dark Heart*, Lt. Col. Anthony Shaffer talks about his capture of a Pakistani Directorate Inter-Services Intelligence female agent who was captured attacking a US outpost in Khowst, Afghanistan. She was found collecting intelligence for the Taliban.[51] Pakistani support for the Taliban and al-Qa'ida needs to end. In addition, the Haqqani Network must be considered as it appears to be ignored by intelligence.

Academic Mahmood Ahmad states that "Today Pakistan is much more unstable than ever before particularly after the surge in drone strikes on its territory under the Obama Administration."[52] This assertion is correct according to every marker that has been used in this chapter. Drone strikes are leading to more suicide terrorist attacks; they are not decreasing terrorism. The Pakistani people are harassed and traumatized by the drone strikes. Innocent people are being killed, as many as six per drone strike. Terrorists are scattering to cities and other populated areas to regroup and escape drone strikes. The effects of American targeted killing are indeed problematic and have created the exact opposite of their original intentions. It is likely that neither Pakistan nor the United States have greatly benefited from American targeted killing in Pakistan.

NOTES

1. Barnett Rubin and Abubakar Siddique, "Resolving the Pakistan-Afghanistan Stalemate" in *Afghanistan from the Cold War through the War on Terror,* ed. Barnett Rubin (Oxford: Oxford University Press, 2013), 360–385.
2. Nasreen Akhtar, "Pakistan, Afghanistan, and the Taliban," *International Journal on World Peace* XXV (4), 2008, 49–73.
3. Alex Strick Van Linschoten and Felix Kuehn, *An Enemy We Created, The Myth of the Taliban-Al Qaeda Merger in Afghanistan* (Oxford: Oxford University Press, 2012).
4. Anatol Lieven, *Pakistan, A Hard Country* (New York: Penguin Books, 2011), 193.
5. Claudio Franco, "The Tehrik-E Taliban Pakistan," in *Decoding the New Taliban, Insights from the Afghan Field,* ed. Antonio Giustozzi (New York: Columbia University Press, 2010), 274.

6. Pervez Musharraf, *In the Line of Fire, A Memoir* (New York: Free Press, 2006), 204–205.

7. Ismail Khan, "Night Raid Kills Nek, Four Other Militants: Wana Operation," *Dawn*, June 19, 2004, Accessed November 19, 2015, http://www.dawn.com/2004/06/19/top1.htm.

8. Dana Priest, "Surveillance Operation in Pakistan Located and Killed Al Qaeda Official," *The Washington Post*, May 15, 2005, Accessed November 19, 2015, http://www.washingtonpost.com/wp-dyn/content/article/2005/05/14/AR2005051401121.html.

9. *Telegraph*, "Al-Qa'ida Number three 'Killed by CIA Spy Plane' in Pakistan," *Telegraph*, December 4, 2005, Accessed November 19, 2015, http://www.telegraph.co.uk/news/worldnews/asia/pakistan/1504718/Al-Qa'ida-number-three-killed-by-CIA-spy-plane-in-Pakistan.html.

10. Jeremy Scahill, "The Secret US War in Pakistan," *The Nation*, November 23, 2009, Accessed November 19, 2015, http://www.thenation.com/article/secret-us-war-pakistan/.

11. Vahid Brown and Don Rassler, *Fountainhead of Jihad, the Haqqani Nexus, 1973–2012* (New York: Columbia University Press, 2013), 153.

12. Vahid Brown and Don Rassler, *Fountainhead of Jihad, the Haqqani Nexus, 1973–2012* (New York: Columbia University Press, 2013), 153.

13. Jon Boone (Islamabad) and Peter Beaumont, "Pervez Musharraf admits permitting 'a few' US drone strikes in Pakistan," *The Guardian*, April 12, 2013, accessed http://www.theguardian.com/world/2013/apr/12/musharraf-admits-permitting-drone-strikes.

14. Jon Boone (Islamabad) and Peter Beaumont, "Pervez Musharraf admits permitting 'a few' US drone strikes in Pakistan," *The Guardian*, April 12, 2013, accessed http://www.theguardian.com/world/2013/apr/12/musharraf-admits-permitting-drone-strikes.

15. AFP, "Zardari calls for an end to drone strikes," *The Express Tribune*, September 15, 2012, Accessed December 2, 2015, http://tribune.com.pk/story/437202/pakistan-sees-us-as-major-development-partner-pm-ashraf/.

16. Sputnik International, "US Drone Strikes in Pakistan Breed Terrorism—President to Sputnik," March 16, 2015, accessed October 5, 2015, http://sputniknews.com/asia/20150316/1019558238.html.

17. Saeed Shah, Adam Entous, and Gordon Lubold, "U.S. Threatens to Withhold Pakistan Aid," *The Wall Street Journa*l, August 21, 2015, accessed October 5, 2015, http://www.wsj.com/articles/u-s-threatens-to-withhold-pakistan-aid-1440163925.

18. *International Crisis Group*, "Drones: Myths and Realities in Pakistan," May 21, 2013, accessed September 28, 2015, http://www.crisisgroup.org/en/regions/asia/south-asia/pakistan/247-drones-myths-and-reality-in-pakistan.aspx.

Salman Masood, "C.I.A. Leaves Base in Pakistan Used for Drone Strikes," *The New York Times*, December 1, 2011, accessed September 24, 2015, http://www.nytimes.com/2011/12/12/world/asia/cia-leaves-pakistan-base-used-for-drone-strikes.html?_r=0.

19. *The Bureau of Investigative Journalism*, "Naming the Dead," accessed September 16, 2014, http://www.thebureauinvestigates.com/namingthedead/the-dead/?lang=en.

For U.S. drone attacks on Pakistan: *The Center for the Study of Targeted Killing*, accessed September 16, 2014, http://www.targetedkilling.org/.

20. *The Bureau of Investigative Journalism*, "Naming the Dead," accessed September 16, 2014, http://www.thebureauinvestigates.com/namingthedead/the-dead/?lang=en.

21. Haider Ali Hussein Mullick, "Power by Proxy: Al-Qa'ida in Pakistan," in *Al-Qa'ida After Ten Years of War A Global Perspective of Successes, Failures, and Prospects*, eds. Norman Cigar and Stephanie E. Kramer (Quantico, Virginia: Marine Corps University Press, 2011), 160–179.

22. Michael Boyle, "The Costs and Consequences of Drone Warfare," *International Affairs*, 89 I, 2013, 3.

23. Palash Ghosh, "I Would Order Air Force To Shoot Down US Drones: Pakistan Opposition Leader Imran Khan," *International Business Times,* October 5, 2012, accessed January 27, 2015, http://www.ibtimes.com/i-would-order-air-force-shoot-down-us-drones-pakistan-opposition-leader-imran-khan-842017.

24. Jon Boone, "US limits Pakistan drone strikes amid political battle over military moves," *The Guardian,* February 5, 2014, accessed February 5, 2014, http://www.theguardian.com/world/2014/feb/05/us-limits-pakistan-drone-strikes.

25. In a hearing before the Committee on Armed Services, Dr. David Kilcullen from the Center for a New American Security stated that the Pakistani people have not been brought into it [drone warfare in Pakistan] and that they should be consulted about drones and targeted killing in Pakistan. House Hearing 111 Congress, "Effective Counterinsurgency: The Future of the U.S.-Pakistan Military Partnership." Hearing before the full committee of the Committee on Armed Services, April 23, 2009, accessed October 15, 2014, http://www.gpo.gov/fdsys/pkg/CHRG-111hhrg52666/content-detail.html.

26. Anatol Lieven, *Pakistan, A Hard Country* (New York: Penguin Books, 2011), 241.

27. Chris Woods, "CIA drone strikes violate Pakistan's sovereignty, says senior diplomat," *The Guardian,* August 3, 2012, accessed January 27, 2015, http://www.theguardian.com/world/2012/aug/03/cia-drone-strikes-violate-pakistan.

28. Saeed Shah, "Anti-Americanism rises in Pakistan over U.S. motives," McClatchy D.C., September 7, 2009, accessed January 27, 2015, http://www.mcclatchydc.com/2009/09/07/74966/anti-americanism-rises-in-pakistan.html.

29. Imtiaz Ali and Massoud Ansari, "Pakistan Fury as CIA Airstrike on Village Kills 18," *Telegraph*, January 15, 2006, accessed September 24, 2015, http://www.telegraph.co.uk/news/worldnews/asia/pakistan/1507895/Pakistan-fury-as-CIA-airstrike-on-village-kills-18.html.

30. Ahmed Rashid, *Pakistan on the Brink* (New York: Penguin, 2012), 54.

31. As cited by Brian Glyn Williams, "The Cia's Covert Predator Drone War in Pakistan, 2004-2010: The History of An Assassination Campaign," *Studies in Conflict and Terrorism* 33, 2010, 871–892.

32. *Pew Research Global Attitudes Project*, "Global Indicators Database: Drone Strikes." 2014, accessed January 27, 2015, http://www.pewglobal.org/database/indicator/52/survey/16/.

33. Medea Benjamin, *Drone Warfare Killing by Remote Control* (New York: Verso, 2013), 108.

34. C. Christine Fair, Karl Kaltenthaler and William Miller, "Pakistani Political Communication and Public Opinion on U.S. Drone Attacks," *Journal of Strategic Studies* 38 (6), 2015.

35. *Pew Research Global Attitudes & Trends*. "A Less Gloomy Mood in Pakistan," August 27, 2014, accessed May 5, 2016, http://www.pewglobal.org/2014/08/27/a-less-gloomy-mood-in-pakistan/.

36. Medea Benjamin, *Drone Warfare Killing by Remote Control* (New York: Verso, 2013), 108.

37. Columbia Law School Human Rights Clinic, The Civilian Impact of Drones: Unexamined Costs, Unanswered Questions, *Center for Civilians in Conflict*, 2012, accessed November 17, 2015 www.civiliansinconflict.org 23–24.

38. Joshua Foust, *American Security Project*, "Understanding the Strategic and Tactical Considerations of Drone Strikes," January 2013, accessed September 25, 2014 www.americansecurityproject.org 10.

39. Rosemarie Skaine, *Women of Afghanistan in the Post-Taliban Era, How Lives Have Changed and Where They Stand Today* (Jefferson, NC: McFarland & Company, Inc., Publishers, 2008), 19.

40. Wali Aslam, "Terrorist Relocation and the Societal Consequences of US Drone Strikes in Pakistan," *The Remote Control Project*, June 2014, accessed October 10, 2014, http://remotecontrolproject.org/wp-content/uploads/2014/09/Wali-Report.pdf.

41. For Pakistan suicide bombings: *Pakistan Body Count.org*, Pakistan Body Count, accessed September 16, 2015, http://pakistanbodycount.org/suicide_bombing.

For U.S. drone attacks on Pakistan: *The Center for the Study of Targeted Killing*, accessed September 16, 2014, http://www.targetedkilling.org/.

42. Leila Hudson, Colin S. Owens, and Matt Flannels, "Drone Warfare: Blowback from the New American Way of War," *Middle East Policy Council*, XVIII, no.3 (Fall 2011).

Michael Boyle, "The Costs and Consequences of Drone Warfare," *International Affairs*, 89 I, 2013.

43. Elisabetta Povoledo, "Giovanni Lo Porto Was Known as Aid Worker Drawn to Needy," *The New York Times,* April 23, 2015, http://www.nytimes.com/2015/04/24/world/europe/american-drone-strike-italian-aid-worker-giovanni-lo-porto.html.

44. Alex Rodriguez, *Los Angeles Times,* "Pakistani Death Squads go after Informants to U.S. Drone Program." December 28, 2011, accessed August 21, 2014. http://articles.latimes.com/2011/dec/28/world/la-fg-pakistan-death-squads-20111228.

And see Joshua Foust, Understanding the Strategic and Tactical Considerations of Drone Strikes, *American Security Project,* January 2013, accessed September 25, 2014 www.americansecurityproject.org, 10.

Columbia Law School Human Rights Clinic, The Civilian Impact of Drones: Unexamined Costs, Unanswered Questions, *Center for Civilians in Conflict*, 2012, accessed November 17, 2015, www.civiliansinconflict.org.

45. Joshua Foust, "Understanding the Strategic and Tactical Considerations of Drone Strikes," *American Security Project,* January 2013, accessed September 25, 2014, www.americansecurityproject.org, 10.

46. Al Arabiya News, "Guidebook on Muslim Spies Reveals al-Qa'ida's Fear," *Al Arabiya News*, July 10, 2009, accessed December 12, 2015, http://www.alarabiya.net/articles/2009/07/10/78352.html.

47. Bill Roggio, "'Good Taliban' leader Mullah Nazir killed in US drone strike," *The Long War Journal*, January 3, 2013, accessed December 2, 2015, http://www.longwarjournal.org/archives/2013/01/good_taliban_leader_1.php.

48. See Brian Glyn Williams, *Predators* for a complete list 2009 militants killed by drone strikes including many unknowns, 94–96.

49. Bill Roggio, US Strikes in Pakistan, HVTs, *The Long War Journal*, 2015, accessed November 17, 2015, http://www.longwarjournal.org/pakistan-strikes-hvts.

50. Avery Plaw and Matthew S. Fricker, "Tracking the Predators: Evaluation the US Drone Campaign in Pakistan," *International Studies Perspectives*, 13, 2012, 344–365.

51. Anthony Shaffer, *Operation Dark Heart: Spycraft and Special Ops on the Frontlines of Afghanistan and the Path to Victory* (New York: Thomas Dunne Books, 2010), 67.

52. Mahmoud Ahmad, "The United States Use of Drones in Pakistan: A Politico-Strategic Analysis," *Asian Affairs: An American Review* 41(1), 2014, 21–29.

Chapter 4

Somalia

Lords of War

THE MODERN HISTORY OF AL-SHABAAB IN SOMALIA

"No one really knows the population of Somalia, but during the past twenty years, somewhere between one third and one half of the six-to-eight million inhabitants have fled their homes. There were over one and a half million refugees abroad."[1] Somalia is literally a land of war, controlled by lords of war and this has created a massive departure of Somalians from Somalia in the last few decades. The primary lords of war are *Harakat al-Shabaab al-Mujahideen* or simply al-Shabaab while the clans of Somalia and their warlords are the secondary bellicose militias. In English "al-Shabaab" translates to "Movement of the Striving Youth." Al-Shabaab, the focal point of American targeted killing in Somalia, grew out of not only a civil war but also an international war in a country that had been at war literally since its independence.

In the beginning, Somalia was considered more likely to be a success story than a failed state. Unlike other nations carved out from the Berlin Conference in 1884, Somalia was one of the most ethnically homogenous countries with over 85 percent of Somalians being from the *Hamitic* ethnic group. In addition, almost all Somalis share the same religion of Sunni Islam and the same languages, either Arabic or Somali.

Despite this prognostication for peace, the history of violence in Somalia for decades is generally based on clan ties. Although most Somalians are of the same ethnicity and religion, societal stratification is based on clans, sub-clans, and sub-sub clans. In order of prestige, there are the *bilis* or noble clans (nomadic pastoralists), the Digil and Rahanweyn (agropastoralists and weapons carrying), and then those who are ethnic minorities working in ritually unclean professions like tanning. Any injury against another person is understood at a group level. The guilty group must make amends to the injured group as each group is held responsible for any action by an individual. Claims produce counterclaims, which produce counterclaims and this leads to constant warfare. Somalian citizenship does not exist but clan identification does. Al-Shabaab, the targeted terrorist group of Somalia, is literally composed of over 70 percent membership of men from weaker clans, particularly the Rahanweyn-Digil or Hawiye clans.[2]

Contradictive to this argument and perhaps even the Somalian people is that since the creation of Somalia, successive presidential administrations have tried to unite the five regions of Somalia into one country including the following regions or territories: British Somalia (currently Somaliland), Italian Somalis (Somalia today), the French territory of Djibouti (achieved independence in 1977), the Ogaden (part of Ethiopia), and the northern province of Kenya. No one has been successful in uniting all Somalians under a single flag. For the purpose of posterity, Somalia's flag even has a five pointed star implying that all the regions of Somalia will be united one day.

Initially, Somalia consisted of British Somaliland and Italian Somaliland after the Berlin Conference in 1884. The Italian occupation lasted until 1941 when the Italians were forced as an Axis power to hand military control over to the British. Northern Somalia (British) remained a protectorate while Southern Somalia became a United Nations Trusteeship until 1949 under the Italians. Somalia gained its independence in July 1960 when both regions joined under the Somali Republic. In 1969, Mohamed Said Barre seized power and formed the Somali Democratic Republic. Although President Barre declared Somalia to be a socialist republic, he formed an alliance with anyone who would fill his coffers with foreign aid. The USSR was Somalia's largest donor until 1987, when they abruptly began sponsoring Ethiopia instead. In turn, Barre turned to the United States for financial assistance. In 1991, the Somalian Republic collapsed as civil war broke out in Somalia, partly due to Said's crony politics in the distribution of government positions to his own clan. What followed was a bloody civil war in which numerous powers including the United Nations, Kenya, Ethiopia, Djibouti, and international organizations were vying for control until 2006.

As a failed state, lacking any central government, the Union of Islamic Courts (UIC) was able to usurp power in Somalia in June 2006 under Abdullahi Yusuf. In reply to the UIC takeover of Somalia, Osama bin Laden published a recorded message on July 1, 2006 stating that that the UIC were "an agent of foreign apostates" and called for Somalians to attack the Yusuf government "quickly to make sure he doesn't flee."[3] In response, the UIC took over vast swaths of territory in southern and central Somalia in June through December 2006.

The UIC had declared victory over The Alliance for the Restoration of Peace and Counter-Terrorism (ARPCT), a group of American-backed warlords, who were more preoccupied with money than their treatment of civilians. President George W. Bush expressed his concern over the triumph of Islamic extremists. Meanwhile, al-Qa'ida warned anyone who would want to destroy the UIC to stay out of Somalia as foreign powers were not welcome. In December 2006, the UIC was defeated by the Transitional Federal Government (TFG) or current government of Somalia[4] with the help of Ethiopia who invaded Somalia in December 2006 with over 14,000 troops. The UIC was officially dealt with and disbanded with several members fleeing the country in early 2007. Some men split from the UIC joining al-Shabaab who had sprung up to defend Somalia from the Ethiopian troops.[5]

The true origins of al-Shabaab are not easy to pin down. In fact, the leadership structure and founding documents are scarce, leading to little academic scholarship on the group. This may be because the group is still relatively new but one would think that the potency of al-Shabaab would call for more literature on the organization. Several stories exist about its founding and all of these stories have some kernel of truth in them. In his recently published book *Al-Shabaab in Somalia*,[6] based on field research in Somalia, Stig Jarle Hansen found that the group's initial origins were based on veterans returning from their jihad in Afghanistan. In addition, al-Shabaab gained some men from the Jamaa'a al-Shabaab (Youth Group), which split from the Union of Islamic Courts (UIC). It is also true that several high-profile meetings finally led to the initial creation of the group, which occurred in 2006.

In its early days in 2007, al-Shabaab derived popular support from its fight against the Ethiopian invaders. American air strikes targeting al-Shabaab and al-Qa'ida put a large toll on the group killing several important leaders. Al-Shabaab managed to survive by using the Internet as a solid recruiting tool to rebuild its forces.[7] In 2008, al-Shabaab made massive territorial gains including most of the south and the port of Kismayo. Their numbers continued to expand as the Ethiopian military was brutal, killing many civilians, providing empathy for the Islamists. The United States tended to back the most brutal anti-Islamists they could find so the United States was not much help either for civilians. The Somalian police were rarely paid so they did not present much of a threat to the group and bribery could solve many problems. Anyone who sold goods to Ethiopians or who worked for the Somalian Government was threatened or killed. In January 2009, Ethiopia gave up, withdrawing its troops, leaving al-Shabaab in control of the capitol Mogadishu and southern Somalia.[8] From that point on Somalians depended on the African Union Mission in Somalia (AMISOM) force of 4500 to protect it and an anti-Islamist group Ahlu Sunna Waljamaca to protect them from al-Shabaab. On January 31, 2009, Sheikh Sharif Ahmed was elected president while al-Shabaab controlled most of southern Somalia.[9]

Al-Shabaab formally joined al-Qa'ida in 2010 along with its 1800 foreign fighters from outside Somalia, although al-Qa'ida has had links with Somalia since the early 1990's when Osama bin Laden resided in Khartoum, Sudan.[10] The exact time that al-Shabaab joined al-Qa'ida varies from scholar to scholar. For example Daniel Byman states that al-Shabaab declared their allegiance to al-Qa'ida in September 2009.[11] Osama bin Laden called al-Shabaab, "one of the most important armies in the Mujahid Islamic Battalion."[12]

The reasons for the large influx of foreign fighters who joined al-Shabaab are interesting. It is speculated that native Somalians were relatively disgusted at the brutality of the group and so al-Shabaab began recruiting from other countries.[13] Somalians have their own practice of cultural Islam and strict Wahhabism does not mesh well with Somalian culture. In addition, many Somalian al-Shabaab fighters were criminals and were looking for some type of protection, fearful of the backlash from their criminal activities. This is perhaps one of the reasons that al-Shabaab members always have their faces covered to preserve anonymity; this practice does not appear in any other Islamic terrorist group.

Children in their teens are frequently abducted by al-Shabaab to serve in the front lines. Refugee camps like Dadaab in Kenya are prime locations used to recruit and abduct young boys.[14] At the time this book was being written, Dadaab was shut down by Kenya as its primary funder the United States was worried about the camp's function as a recruitment base for al-Shabaab. In addition to pressure from the United States for more Kenyan oversight of the camp, Kenya has been attacked by al-Shabaab. The Westgate Mall terrorist attack occurred in 2013 where around seventy people were killed and hundreds were wounded. After the closure of Dadaab, approximately 330,000 Somalian people are expected to return to Somalia and a situation they had fled in terror from initially. The Foreign Affairs Minister of Kenya Amina Mohamed said, "[It is] an issue of balancing national security interests and international obligations; I've been doing what I am doing for over thirty years. I've never come across a country that puts its international obligations above its national interests."[15]

From 2010 to 2011, al-Shabaab did not have the best years due to a national drought, in-fighting within the organization, and increasing involvement in piracy. By 2012, al-Shabaab had managed to increase its territory again. In 2013, leader Ahmed 'Godane' killed, sidelined, or imprisoned other leaders within al-Shabaab including American Omar Hammami, known as al-Amriki, and his former mentor Ibrahim al-Afghani. This leadership crisis turned out the best for al-Shabaab as most internal dissent within the organization was squashed. One of the most deadly al-Shabaab attacks occurred in June 2013, when the group attacked a United Nations compound in Mogadishu killing twenty-two people. In February 2014, the group attacked the presidential palace with a car bomb and armed assailants, which killed twelve people.

The brutality of al-Shabaab is alarming. The group will cross amputate the arms and legs of thieves. Al-Shabaab also stoned a thirteen-year-old girl to death who had been gang raped, stating she had committed adultery. In society, men without beards are whipped and young girls who cannot afford hijab are sent to prison. Women who wear traditional Somalian dress are whipped. Improper haircuts will be corrected immediately on the street. Young girls in Shabaab-held areas are forced into marriage with al-Shabaab militants or future suicide bombers as rewards for their impending sacrifices.[16] Teachers are whipped for teaching girls. Any person who receives medical treatment from the infidels in the war-torn society would be considered unclean and would have their tongue cut out.[17] Sufi Muslims are harassed, killed, and their places of worship are destroyed. Inside the organization of al-Shabbab is the *Amniyat*, who function as internal secret police, punishing anyone who denies or repudiates the organization. Some Somalian citizens appreciate al-Shabaab as they have stamped out lawlessness in certain areas, establishing their own system of Sharia courts. The rape and theft went down tremendously after the group took power in Mogadishu although barbaric punishments are still imposed upon people.[18] Al-Shabaab's power rests on a climate of fear and retaliation, even within their own organization.

It is believed that al-Shabaab makes about 35 million to 50 million per year in port revenues. At least fifteen million dollars of that money is based on trade in charcoal and sugar.[19] Al-Shabaab has mandated that it is illegal to sell charcoal as it is destructive to the Somalian environment, yet al-Shabaab continues to make large profits off the trade of charcoal.

THE EFFECTS OF AMERICAN DRONE STRIKES ON SOMALIA

The history of counterterrorism efforts in Somalia by the United States have never been highly successful from a policy perspective even though the United States has been carrying out extensive covert military actions in Somalia since 2001, including having at least three JSOC members on the ground at all times in Somalia.[20] On December 13, 2002, the Bush Administration officially opened Camp Lemonnier in Djibouti next to the international airport. Originally a French munitions storage facility, Camp Lemonnier was converted into an American Naval base. Camp Lemonnier is officially a United States' Naval Base although it is used to conduct counterterrorism multi-force operations in both Yemen and Somalia. It is also the primary base of operations for United States Africa Command and is the only permanent base in Africa.[21]

American counterterrorism efforts to defeat al-Qa'ida in Somalia began on the ground shortly after the opening of Camp Lemonnier. On January, 2003, Mohamed Afrah Qanyare stood on his tarmac waiting for a group of American intelligence personnel to land; he had recently met a few Americans in Nairobi, Kenya where he was told that the CIA wanted his help. Qanyare was a local warlord in Somalia who controlled area to the north of Mogadishu and had made a small fortune controlling an airport, which primarily brought in shipments of *mira* or khat, an addictive plant leaf chewed in the Horn of Africa. The CIA primarily wanted Qanyare to locate al-Qa'ida members in Somalia and East Africa. On average, the Americans paid Qanyare 100,000 to 150,000 per month to use his services and his airport. This arrangement eventually became known as the "Alliance for the Restoration of Peace and Counterterrorism" or "Operation Black Hawk" in the intelligence community.

Efforts to locate and destroy al-Qa'ida members led to the increased financial support of warlords throughout Somalia. Somalia's foreign minister, Ismail Mahmoud "Buubaa" Hurre stated after Bush's "Either you are with us or you are with the terrorists" speech that he penned a letter to the United States that he was on board. The United States was not really responsive. Buubaa found that instead of strengthening the Somalia government, "they [CIA] started cooperating with the warlords, thinking that the best way to combat terrorism was to help the warlords become stronger, and chase away the fundamentalists from Somalia. That backfired."[22]

According to *The Bureau of Investigative Journalism*, the United States had Somalian warlords capture seven or eight al Qa'ida members, taking them from Somalia. These members include Suleiman Abdallah, who was taken from a hospital in Mogadishu in March 2003 and was held and tortured over five years by the United States. Mohammed Ali Isse was also taken and was held on a ship off the coast of Djibouti until he was transferred to a prison in Somaliland.[23]

Problematically numerous warlords funded by the United States were also torturing, killing, or kidnapping innocent people. In his book *Dirty Wars*, Jeremy Scahill quotes Abdirahman "Aynte" Ali, a Somalian expert who stated,

> They were either capturing them, and then renditioning them to Djibouti, where there is a major American base. Or in many cases, they were chopping their head off and taking the head to the Americans and telling them, 'We killed this guy.' The vast majority of people they killed had nothing to do with the War on Terror.[24]

Scahill also quotes a Somalian militia leader who had worked with the Americans in counterterrorism operations, "As many as seventeen suspected terrorists have been apprehended in Mogadishu alone since 2003-all but three are apparently innocent."[25]

Two sites in Somalia are used as detention centers in Somalia. The first of these is a compound located at Mogadishu's Aden Adele International Airport run by the CIA. The CIA has its own aircraft hangers at the airport to conduct operations guided by the Somalian military. There is also a training program for Somalian intelligence agents at the compound who receive approximately 200 American dollars salary per month from the United States. The second location is a secret prison that is buried in the basement of the Somalian National Security Agency headquarters in Mogadishu. Al-Shabaab and al-Qa'ida militants in Somalia and Kenya are interrogated here. The prison is filthy and infested with bed bugs and mosquitoes.[26]

Although they were the identifiers of al-Shabaab militants on the ground for extraordinary rendition and torture, warlords were also used to locate militants for drone strikes. From 2011 to the end of 2015, the United States has conducted approximately nineteen drone strikes in Somalia. As can be seen in Figure 4.1: *American Drone Strikes in Somalia 2011–2015*,[27] these strikes began to increase tremendously at the end of 2015. This trend has continued to increase into 2016 and it is suspected that drone strikes in Somalia will further continue to increase if ISIS moves into Africa and/or if al-Shabaab continues to grow. The al-Qa'ida ties of al-Shabaab have made it a prime target for continued American drone strikes.

On October 26, 2012, the US military confirmed for the first time that armed drones fly out of Camp Lemonnier in Djibouti. Currently Camp

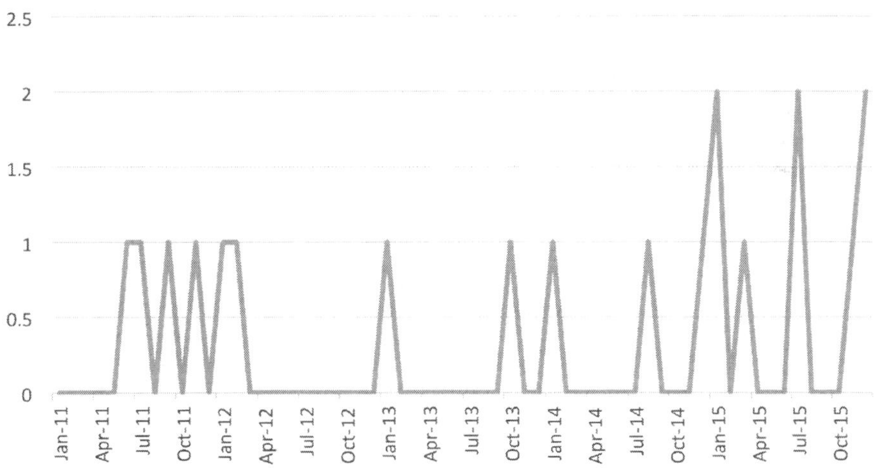

Figure 4.1 American Drone Strikes in Somalia 2011–2015. *Source*: *The Bureau of Investigative Journalism*, "US Strikes in Somalia: 2007–Present," 2016, accessed May 15, 2016, https://docs.google.com/spreadsheets/d/1-LT5TVBMy1Rj2WH30xQG9nqr8-RXFVvzJE_47NlpeSY/edit#gid=859698683.

Lemonnier is the 'the busiest Predator drone base outside of Afghan war zone'. Over 300 JSOC personnel coordinate strikes and around sixteen drones operate on a daily basis from the Camp. The drones are operated and maintained by the eighty-seven-member 60th Air Force Expeditionary Reconnaissance Squadron, nicknamed the "East Africa Air Pirates." In August 2012, the US military told Congress it wanted to spend $1.4 billion to increase the capacity of the Camp and to add 1000 more JSOC personnel to the already 3200 people at the Camp.[28]

Many of the United States drone strikes in Somalia that fly from Camp Lemmonier, operated by Air Force pilots in the United States (mainly Creech), have focused on al-Shabaab training camps. On June 23, 2011, Predators struck a militant training camp, south of Kismayo. The strikes were aimed at two senior militants who were reportedly planning an imminent terrorist attack on the UK, including Ibrahim al Afghani.[29] After the strike a JSOC team in helicopters came to collect the bodies.[30] On July 6, 2011, American drones or planes reportedly hit three al-Shabaab militant training camps in Afmadow.[31] On March 10, 2015, "unidentified armed drones" targeted two alleged al-Shabaab camps in the towns of Torato and Ambereso; the number of people killed is unknown.[32] In March 2016, the United States targeted the Raso training camp, where over 150 fighters were killed in an al-Shabaab graduation ceremony; this strike is the deadliest US drone strike ever.[33] On April 11–12, 2016, drone strikes were sent to destroy another camp in southern Somalia, killing about six people.[34]

So far, the United States has killed few major al-Shabaab leaders in Somalia with drone strikes. On September 1, 2014, Ahmed Abdi Godane, the Emir of al-Shabaab was killed in a coordinated air and drone strike along with six other militants.[35] He was quickly replaced with Abu Ubaidah. External Operation Chief Yusuf Dheeq was killed in February 2016 outside of Mogadishu.[36] His replacement Adan Garar, a senior member of al-Shabaab secret police, the Amniyat, was killed on March 12, 2015 by a drone strike. On March 31, 2016 Hassan Ali Dhoore was killed in a strike south of Jilib with two other militants; Dhoore helped to plan the Christmas Day 2014 attack at the Mogadishu Airport and a March 2015 attack at the Maka al-Mukarramah hotel.[37] On December 2, 2015, a US air strike killed Abirahman Sandhere, a senior al-Shabaab fighter, along with two unidentified associates. On December 22, 2015, Abu Ubaidah, the successor of Godane was killed by American drone strike.

In Figure 4.2, *Deaths from US Drone Strikes Somalia, 2011–2015*,[38] and the minimum and maximum numbers of deaths caused by American drone strikes in Somalia are indicated. Compared with other countries in this books such as Yemen, Pakistan, or Afghanistan, civilian casualties are lower per drone strike. From 2011–2015, there were approximately 6 people killed per drone strike, including the intended targets, the number of which is unknown. According to numbers published by *The Bureau of Investigative Journalism*, this number has more than doubled with the inclusion of half of 2016. With the first half of the year 2016 included, the maximum number of people killed per drone strike is over thirteen people, again including terrorists. This is a testament to the expansion of drone strikes in Somalia with the passage of time and another indicator of what is to be expected to come in the future of Somalia.[39] There are a few mistakes such as the killing of civilians by a drone strike on October 22, 2011 where eleven civilians were killed and as many as twenty civilians were injured. The strike occurred in Afmadow town in the Lower Jubba region.[40] However, there appears to be far less civilian deaths in Somalia than other countries in the drone war.

One of the major questions in this book is whether the drone strikes as part of a large counterterrorism operation in the War on Terror are successful in each individual country. In Somalia, it is still too early to make any educated conclusions concerning the effects of these drone strikes. Somalian public

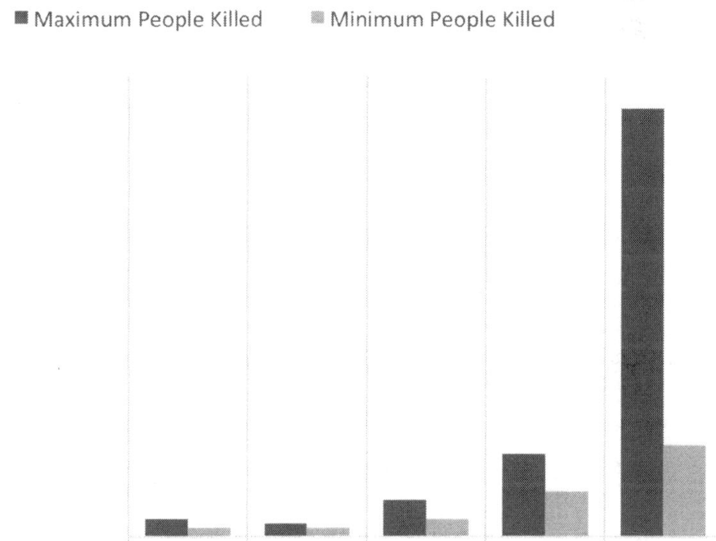

	2011	2012	2013	2014	2015
Maximum People Killed	4	3	8	18	93
Minimum People Killed	2	2	4	10	20

Figure 4.2 Deaths from US Drone Strikes, Somalia, 2011–2015. *Source*: *The Bureau of Investigative Journalism*, "Somalia: Minimum and Maximum People Killed by US Drone Strikes, 2007–present," 2016, accessed May 15, 2016, https://docs.google.com/spreadsheets/d/1-LT5TVBMy1Rj2WH30xQG9nqr8-RXFVvzJE_47NlpeSY/edit#gid=1531042213.

opinion polling concerning civilian opinion on American drone strikes in Somalia is not available, most likely because journalists and polling agencies are hesitant to enter such a dangerous country. Although civilian casualties are less and drone strikes are much less than in other countries such as Afghanistan, Pakistan, or Yemen, terrorist attacks in Somalia have not decreased. In Figure 4.3: *Al-Shabaab Terrorist Attacks in Somalia, 2007–2014*,[41] the number of terrorist attacks committed by al-Shabaab from their initial creation until 2014 is displayed. Although there is not enough data to quantitatively test the relationship between drone strikes and al-Shabaab terrorist attacks, it is evident that terrorist attacks have increased with the increase of American drone strikes. The number of terrorist attacks committed just by al-Shabaab[42] have increased tremendously since the drone strikes began in 2007. It is clearly evident that drone strikes are not decreasing terrorist attacks in Somalia, but in fact may be contributing to their increase.

Like other countries that endure American drone strikes, Somalia has not publicly given their consent for American drone strikes to occur in their country. Former President Sharif stated after being asked if drone strikes weaken or strengthen his government, "Both at the same time. For our sovereignty, it's not good to attack a sovereign country. That's the negative part. The positive part is you're targeting individuals who are criminals."[43] Current President Hassan Sheikh Mohamud said in a note on Twitter on April 31 "that his

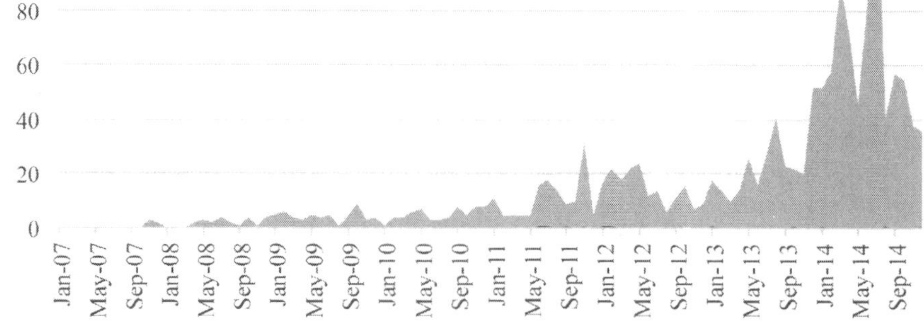

Figure 4.3 Al-Shabaab Terrorist Attacks in Somalia, 2007–2014. *Source*: National Consortium for the Study of Terrorism and Responses to Terrorism, "Global Terrorism Database, Somalia" 2015, accessed May 17, 2016, https://www.start.umd.edu/gtd/search/Results.aspx?chart=overtime&casualties_type=&casualties_max=&country=182.

country's army and allied forces were in the midst of a heavy attack against al-Shabaab." This implies that President Mohamud has given some kind of consent to the United States although probably not the free reign to conduct strikes at will.[44]

THE DETERMINATION OF AL-SHABAAB

Al-Shabaab is a resilient terrorist organization. Although their numbers are small, most likely around 12,000, the group has not yet been defeated by over fifteen years of American, Ethiopian, Kenyan, and Somalian Government counterterrorism efforts. If anything, the group has grown in size and control of territory. There are a few scholars who curiously enough have postulated why al-Shabaab is so difficult to kill. Professor Hussein Solomon from University of the Free State argues that there are five reasons that al-Shabaab has continued to survive. (1) Al-Shabaab has embraced asymmetric warfare as their tactic of choice, aware of the fact that they cannot defeat AMISOM on the battlefield. (2) Al-Shabaab has consolidated their leadership and ended the power rivalry. Mohamud Godane has created a Special Operations Unit that is responsible for suicide attacks, etc. Its 5000 member force trains daily like a regular military. (3) Al-Shabaab has been successful in exploiting the errors of their foes. For example U.S drone strikes have resulted in more support for al-Shabaab as a result of numerous civilian deaths. Ugandan troops have been found guilty of extorting bribes from Somalians. The propaganda machine of al-Shabaab is powerful and effective. (4) Ethiopia and Kenya have tried to divide Somalians by supporting various warlords, which has not endeared outsiders to the Somalians. (5) Al-Shabaab has managed to find funding despite numerous attempts to stop their revenue stream. Somali pirates,[45] UN Security Council food aid, and Britain's foreign aid have all managed to find their way into al-Shabaab's coffers.[46]

Academic Stig Jarle Hansen adds that al-Shabaab is still a formidable threat for four reasons. The first reason is that al-Shabaab's *Amniyat* or internal secret police is feared among the top leaders of al-Shabaab so followers are less likely to defect. Second, al-Shabaab controls large amounts of territory and is quite wealthy as a result of taxes and piracy. Thirdly, foreign fighters within the organization have shared memories of jihad in Afghanistan etc. and these relationships override any internal disagreements within the organization. Fourth and the last, if some al-Shabaab leaders leave and join the Transitional Federal Government (TFG), the most extreme leaders would likely remain in the organization.[47]

In addition to the previous reasons, if one compares al-Shabaab to ISIS, the results are troublesome. Like ISIS, al-Shabaab has ties with al-Qa'ida and has declared their allegiance to al-Qa'ida. Al-Shabaab has similar views as ISIS including the establishment of the Islamic State and the Caliphate. Al-Shabaab like ISIS are Sunni Muslims with Wahhabi roots.[48] And list goes on. It is not inconceivable to think that ISIS in its current form, controlling parts of Syria and Iraq could unite with al-Shabaab to increase the territory of the Islamic State, thus al-Shabaab would join ISIS and run their part of the Caliphate. Although one might think that al-Shabaab would not give up their autonomy and territory that easily, this was not a big issue for ISIS when Syrian and Iraqi elements were combined. A small power struggle occurred, a few people were killed and the state al-Qa'ida offshoots were united.

ISIS is already a serious problem in the Middle East and despite the big words of the United States' Government, ISIS is not losing. The question is really, is the United States actually winning the war with ISIS? The answer to this question is no, not really. If ISIS and al-Shabaab unite, Africa will be in even greater danger and the establishment of the terrorist al-Qa'ida caliphate will be even harder to demolish, perhaps even requiring *boots on the ground* for the next president of the United States. In terms of growth, al-Shabaab has managed to threaten Somaliland and Puntland in addition to Kenya.[49] Kenya, which has been attacked several times by al-Shabaab has created a home guard of sorts to protect their communities on the border from al-Shabaab.[50]

And yet, al-Shabaab has never attacked the American homeland nor has it specifically attacked the United States before 9–11. In fact, until the United States started attacking al-Shabaab, which occurred directly after September 11, 2001. Al-Shabaab has not concentrated on much outside of the surrounding countries of Somalia. Although the United States began attacking al-Shabaab in 2001, it declared the group a terrorist organization in 2008,[51] simply due to its Islamic extremism. Depending on the scholar, al-Shabaab did not even declare allegiance to al-Qa'ida until after 2010, even as late as 2012 according to the United Nations.[52] It appears that al-Shabaab's only sin before it declared its allegiance to al-Qa'ida was simply being an extreme Islamic organization. This certainly bears witness to the idea that the War on Terrorism is a religious war. The author is not trying to excuse any of the heinous crimes of al-Shabaab but only stating that al-Shabaab never started any conflict with the United States and the United States has been pursuing al-Shabaab since 2001.

CONCLUSION

Hassan Sheikh Mohamud was elected president on September 10, 2012 and is currently the president of Somalia. He was recently named one of *Time*'s most influential people of the year for his efforts on reducing corruption, improving the economy, and creating security reforms. However, Somalia is not improving. The Transitional National Government (TNG) is working in unison with AMISOM to defend against al-Shabaab. They have experienced over 800 deaths since they were first deployed in 2007 and have advanced two kilometers into Mogadishu, which is still held by al-Shabaab. The current number of parties involved and interested in the failed state/civil war is dizzying to keep track of and the conflict has not deescalated. War has decimated the country and turned its houses into bullet-barraged shells of once domestic hearths.[53]

Moreover former President Shairf asked the United States to help train the Somalian military and to focus on creating viable civilian institutions in Somalia. American drone strikes have targeted the rural areas just outside of Mogadishu but have not done much else in the rest of Somalia. If al-Shabaab is not restricted from conquering the rest of Somalia, the military is incapable of stopping it if it continues to grow.[54] The United States continues to fight al-Shabaab on the ground by funding the AMISOM forces who are relatively small and do not conduct the surgical missions necessary to stop al-Shabaab. Instead, AMISOM has managed to abuse the civilian population by indiscriminately targeting al-Shabaab areas in the past, killing any civilians in its path. However, requests were denied when the Pentagon asked for forty-five million to fund AMISOM with more equipment. In yet another testament to Obama's fear of *boots on the ground*, the United States has not been able to conduct any successful targeted mission in al-Shabaab-controlled areas in Mogadishu.[55]

In addition, Somalia contains Camp Lemmonier, just outside of Djibouti, which is the Air Force base mainly responsible for launching drone strikes in the Middle East. At Camp Lemmonier, "The number of takeoffs and landings each month has more than doubled, reaching a peak of 1,666 in July compared with a monthly average of 768 two years ago, according to air-traffic statistics disclosed in Defense Department contracting documents."[56] In September 2012, the Department of Defense awarded a $7 million contract to retrain air-traffic controllers at Ambouli International Airport and improve their English skills. These air-traffic controllers control all aircraft including unmanned aerial vehicle traffic. From January 2011 to October 2012, five Predators have crashed, thus encouraging the need for better air-traffic control in an increasingly busier environment.[57] In neighboring Djibouti, on July 25, 2012, the United Nations published a report that there were so many drones flying around in Somalia that an international crash incident was likely to happen. According to the UN, all of these drones are used for military purposes violating the arms embargo against Somalia.[58] The number of drones flying in Somalia can only grow as there appears to be no end in sight in the increase of drone warfare and the virility of al-Shabaab in Somalia.

In conclusion, although the number of drone strikes in Somalia are low in comparison to other countries, civilian deaths are relatively low also. However, since the United States has increased their number of drone strikes so has al-Shabaab increased their terrorist attacks. Although this relationship cannot be tested, it can be said that drone strikes are not decreasing terrorist attacks or ending terrorism in Somalia. It is possible that drone strikes are contributing to the rise in terrorist attacks. In addition, there is not much research that has been completed on the effects of drone strikes on civilians.

The United States appears content to launch a few drone strikes to kill al-Shabaab leaders and destroy training camps. But someone always steps into the leader's place and al-Shabaab does not appear to be low on recruits, most likely because young Islamic men are eager to play soldier, pursue their jihad, and to die for the glory of Allah. Unless the United States is serious about stopping terrorism in Somalia and protecting the United States from al-Shabaab, little else will occur in Somalia as a result of drone strikes. Somalians need a government that is strong and state that has not failed. Until the standard of living is higher and the warfare stops, shooting a few drone strikes in Somalia has never solved any problems for Somalians and perhaps may make more trouble for the Americans if al-Shabaab choses to fight back in the United States. Al-Shabaab will continue to dominate Somalia until there is a better option of government.

In a recent article published in March 2016 in *The Daily Beast*,[59] the authors state that al-Shabaab is experiencing some of the most successful months in its history. The group is conducting urban bombings and military raids on African Union Mission to Somalia (AMISOM) on a weekly basis. It is a highly organized group that is growing in membership and was the first to establish an Islamic State and have a social media presence. Recruitment efforts in Kenya have been highly successful and foreign fighters are still flocking to Somalia. "This organization still poses a very real international terrorist threat."[60]

NOTES

1. Ben Rawlence, *City of Thorns, Nine Lives in the World's Largest Refugee Camp* (New York: Picador, 2014), 11.
2. Ioannis Mantzikos, "Somalia and Yemen: The Links between Terrorism and State Failure," *Digest of Middle East Studies* 20:2 (2011): 242–260.
3. Shaul Shay, *Somalia between Jihad and Restoration* (New Brunswick: Transaction Publishers, 2008). 31–32.
4. Shaul Shay, *Somalia between Jihad and Restoration* (New Brunswick: Transaction Publishers, 2008).
5. Mohamed Ibrahim, "Somalia and Global Terrorism: A Growing Connection," *Journal of Contemporary African Studies*, 28:3 (2010), 283–295.

6. Stig Jarle Hansen, *Al-Shabaab in Somalia* (London: Hurst and Company, 2016).

7. Stig Jarle Hansen, *Al-Shabaab in Somalia* (London: Hurst and Company, 2016).

8. Hussein Solomon, *Terrorism and Counter-terrorism in Africa* (New York: Palgrave Macmillan, 2015), Chapter 3.

9. Martin N. Murphy, *Somalia: The New Barbary?* (New York: Columbia University Press, 2011), Chapter 26.

10. Matt Bryden, "No Quick Fixes: Coming to Terms with Terrorism, Islam, and Statelessness in Somalia," *The Journal of Conflict Studies* (Fall 2003).

11. Daniel Byman, *Al Qaeda, The Islamic State, and The Global Jihadist Movement, What Everyone Needs to Know* (Oxford: Oxford University Press, 2015), 146.

12. Martin N. Murphy, *Somalia: The New Barbary?* (New York: Columbia University Press, 2011), 152.

13. Martin N. Murphy, *Somalia: The New Barbary?* (New York: Columbia University Press, 2011).

14. Ben Rawlence, *City of Thorns, Nine Lives in the World's Largest Refugee Camp* (New York: Picador, 2014).

15. Murithi Mutiga, "Refugees urge Kenyan leaders to rethink closure of Dadaab camp," *The Guardian*, May 13, 2016, accessed May 13, 2016, http://www.theguardian.com/global-development/2016/may/13/refugees-urge-kenyan-leaders-to-rethink-closure-of-dadaab-camp.

16. Monitoring Group on Somalia and Eritrea, "Report on the Monitoring Group on Somalia and Eritrea pursuant to the Security Council Resolution 2002 (2011)," *United Nations Security Council*, 2011, Accessed May 20, 2016, http://www.un.org/ga/search/view_doc.asp?symbol=S/2012/544.

17. James Fergusson, *The World's Most Dangerous Place* (Philadelphia: Da Capo Press, 2013).

18. Human Rights Watch, *Harsh War, Harsh Peace, Abuses by al-Shabaab, the Transitional Federal Government, and AMISOM* (New York: Human Rights Watch, 2010).

19. Jeremy Scahill, *Dirty Wars: The World is a Battlefield* (New York: Nation Books, 2013), 394.

20. Drones Team, "Somalia: reported US covert actions 2001–2016," *The Bureau of Investigative Journalism*, February 22, 2012, Accessed May 20, 2016, https://www.thebureauinvestigates.com/2012/02/22/get-the-data-somalias-hidden-war/#SOM023.

21. *United States Navy*, "CNIC Camp Lemonnier Djibouti," 2016, Accessed May 18, 2016, http://cnic.navy.mil/regions/cnreurafswa/installations/camp_lemonnier_djibouti.html.

22. Jeremy Scahill, *Dirty Wars: The World is a Battlefield* (New York: Nation Books, 2013), 121.

23. Drones Team, "Somalia: reported US covert actions 2001–2016," *The Bureau of Investigative Journalism*, February 22, 2012, Accessed May 20, 2016, https://www.thebureauinvestigates.com/2012/02/22/get-the-data-somalias-hidden-war/#SOM023.

24. Jeremy Scahill, *Dirty Wars: The World is a Battlefield* (New York: Nation Books, 2013), 128.

25. Jeremy Scahill, *Dirty Wars: The World is a Battlefield* (New York: Nation Books, 2013), 128.

26. Jeremy Scahill, "The CIA's Secret Sites in Somalia," *The Nation*, December 10, 2014, accessed May 20, 2016, http://www.thenation.com/article/cias-secret-sites-somalia/.

27. *The Bureau of Investigative Journalism*, "US Strikes in Somalia: 2007–Present," 2016, accessed May 15, 2016, https://docs.google.com/spreadsheets/d/1-LT5TVBMy1Rj2WH30xQG9nqr8-RXFVvzJE_47NlpeSY/edit#gid=859698683.

28. Phil Stewart, "U.S. strikes al Shabaab training camp in Somalia, more than 150 killed," *Reuters*, March 8, 2016, accessed May 19, 2016, http://www.reuters.com/article/us-usa-somalia-dronestrike-idUSKCN0W91XW.

29. Drones Team, "Somalia: reported US covert actions 2001–2016," *The Bureau of Investigative Journalism*, February 22, 2012, Accessed May 20, 2016, https://www.thebureauinvestigates.com/2012/02/22/get-the-data-somalias-hidden-war/#SOM023.

30. Jeremy Scahill, "The CIA's Secret Sites in Somalia," *The Nation*, December 10, 2014, accessed May 20, 2016, http://www.thenation.com/article/cias-secret-sites-somalia/.

31. Drones Team, "Somalia: reported US covert actions 2001–2016," *The Bureau of Investigative Journalism*, February 22, 2012, Accessed May 20, 2016, https://www.thebureauinvestigates.com/2012/02/22/get-the-data-somalias-hidden-war/#SOM023.

32. Drones Team, "Somalia: reported US covert actions 2001–2016," *The Bureau of Investigative Journalism*, February 22, 2012, Accessed May 20, 2016, https://www.thebureauinvestigates.com/2012/02/22/get-the-data-somalias-hidden-war/#SOM023.

33. Phil Stewart, "U.S. strikes al Shabaab training camp in Somalia, more than 150 killed," *Reuters*, March 8, 2016, accessed May 19, 2016, http://www.reuters.com/article/us-usa-somalia-dronestrike-idUSKCN0W91XW.

34. Drones Team, "Somalia: reported US covert actions 2001–2016," *The Bureau of Investigative Journalism*, February 22, 2012, Accessed May 20, 2016, https://www.thebureauinvestigates.com/2012/02/22/get-the-data-somalias-hidden-war/#SOM023.

35. Drones Team, "Somalia: reported US covert actions 2001–2016," *The Bureau of Investigative Journalism*, February 22, 2012, Accessed May 20, 2016, https://www.thebureauinvestigates.com/2012/02/22/get-the-data-somalias-hidden-war/#SOM023.

36. *DW*, "US targets al-Shabab leader in Somalia with drone strike," April 2, 2016, accessed May 19, 2016, http://www.dw.com/en/us-targets-al-shabab-leader-in-somalia-with-drone-strike/a-18232950.

37. Spencer Ackerman and agencies, "US drone strike kills key al-Shabaab leader in Somalia," *The Guardian*, April 1, 2016, accessed May 19, 2016, http://www.theguardian.com/world/2016/apr/01/us-military-somalia-airstrike-al-shabaab-terrorist-targets.

38. *The Bureau of Investigative Journalism*, "Somalia: Minimum and Maximum People Killed by US Drone Strikes, 2007–present," 2016, accessed May 15, 2016, https://docs.google.com/spreadsheets/d/1-LT5TVBMy1Rj2WH30xQG9nqr8-RXFVvzJE_47NlpeSY/edit#gid=1531042213.

39. *The Bureau of Investigative Journalism*, "Somalia: Minimum and Maximum People Killed by US Drone Strikes, 2007–present," 2016, accessed May 15, 2016, https://docs.google.com/spreadsheets/d/1-LT5TVBMy1Rj2WH30xQG9nqr8-RXFVvzJE_47NlpeSY/edit#gid=1531042213.

40. Drones Team, "Somalia: reported US covert actions 2001–2016," *The Bureau of Investigative Journalism*, February 22, 2012, Accessed May 20, 2016, https://www.thebureauinvestigates.com/2012/02/22/get-the-data-somalias-hidden-war/#SOM023.

41. *National Consortium for the Study of Terrorism and Responses to Terrorism*, "Global Terrorism Database, Somalia," 2015, accessed May 17, 2016, https://www.start.umd.edu/gtd/search/Results.aspx?chart=overtime&casualties_type=&casualties_max=&country=182.

42. These numbers do not include terrorist attacks committed by unknown groups or other terrorist groups in Somalia. The number of terrorist attacks would increase if every terrorist attack in Somalia was included.

43. Jeremy Scahill, "The CIA's Secret Sites in Somalia," *The Nation*, December 10, 2014, accessed May 20, 2016, http://www.thenation.com/article/cias-secret-sites-somalia/.

44. Harun Maruf and Jeff Seldin, "Shootings in Somalia's Capital Leave 7 Dead," *Voice of America*, April 2, 2016, accessed May 23, 2016, http://www.voanews.com/content/us-drone-attack-kills-top-al-shaba-assassin-somalia/3265379.html.

45. Martin N. Murphy, *Somalia: The New Barbary?* (New York: Columbia University Press, 2011).

46. Hussein Solomon, *Terrorism and Counter-terrorism in Africa* (New York: Palgrave Macmillan, 2015), Chapter 3.

47. Stig Jarle Hansen, *Al-Shabaab in Somalia* (London: Hurst and Company, 2016), Conclusions.

48. Karen Armstrong, "Wahhabism to ISIS: How Saudi Arabia Exported the Main Source of Global Terrorism," *The New Statesman*, November 27, 2014, accessed May 20, 2016, http://www.newstatesman.com/world-affairs/2014/11/wahhabism-isis-how-saudi-arabia-exported-main-source-global-terrorism.

49. Denise Riviera, "Harmony and Mayhem in Somalia," *The Advocate*, May 6, 2016, accessed May 6, 2016, http://gcadvocate.com/2016/05/06/harmony-mayhem-somalia/.

50. Will Swanson and Ariel Zirulnick, "Hiking the 'Rat Tracks,' Hunting for al-Shabaab," *Foreign Policy*, May 20, 2016, accessed May 20, 2016, http://foreignpolicy.com/2016/05/20/hiking-the-rat-tracks-hunting-for-al-shabab/?utm_source=Sailthru&utm_medium=email&utm_campaign=New%20Campaign&utm_term=%2AEditors%20Picks.

51. Counterterrorism Guide, "Al-Shabaab," *National Counterterrorism Center*, 2016, accessed May 20, 2016, https://www.nctc.gov/site/groups/al_shabaab.html.

52. Monitoring Group on Somalia and Eritrea, "Report on the Monitoring Group on Somalia and Eritrea pursuant to the Security Council Resolution 2002 (2011),"

United Nations Security Council, 2011, Accessed May 20, 2016, http://www.un.org/ga/search/view_doc.asp?symbol=S/2012/544.

53. James Fergusson, *The World's Most Dangerous Place* (Philadelphia: Da Capo Press, 2013).

54. Jeremy Scahill, "The CIA's Secret Sites in Somalia," *The Nation*, December 10, 2014, accessed May 20, 2016, http://www.thenation.com/article/cias-secret-sites-somalia/.

55. Jeremy Scahill, "The CIA's Secret Sites in Somalia," *The Nation*, December 10, 2014, accessed May 20, 2016, http://www.thenation.com/article/cias-secret-sites-somalia/.

56. Craig Whitlock, "Remote U.S. Base at Core of Secret Operations," *The Washington Post*, October 25, 2012, accessed May 20, 2016, https://www.washingtonpost.com/world/national-security/remote-us-base-at-core-of-secret-operations/2012/10/25/a26a9392-197a-11e2-bd10-5ff056538b7c_story.html.

57. Craig Whitlock, "Remote U.S. Base at Core of Secret Operations," *The Washington Post*, October 25, 2012, accessed May 20, 2016, https://www.washingtonpost.com/world/national-security/remote-us-base-at-core-of-secret-operations/2012/10/25/a26a9392-197a-11e2-bd10-5ff056538b7c_story.html.

58. Monitoring Group on Somalia and Eritrea, "Report on the Monitoring Group on Somalia and Eritrea pursuant to the Security Council Resolution 2002 (2011)," *United Nations Security Council*, 2011, Accessed May 20, 2016, http://www.un.org/ga/search/view_doc.asp?symbol=S/2012/544.

59. The Other Jihadis, "The Return of al-Shabaab," *The Daily Beast*, March 15, 2016, accessed May 25, 2016, http://www.thedailybeast.com/articles/2016/03/15/the-return-of-al-shabaab.html.

60. The Other Jihadis, "The Return of al-Shabaab," *The Daily Beast*, March 15, 2016, accessed May 25, 2016. http://www.thedailybeast.com/articles/2016/03/15/the-return-of-al-shabaab.html.

Chapter 5

Libya

The Elimination of an Eccentric Dictator

The former eccentric dictator, Muammar Qaddafi[1] came to power in September 1969 after leading a bloodless military coup against Crown Prince Hasan al-Rida. Initially called the White Revolution, it later became known as the September Revolution. After the September Revolution ended, Qaddafi became the de facto leader of Libya with the rank of Colonel.[2] In 1973, Qaddafi instituted his plan: the four pillars of socialism which consisted of: socialism, popular democracy, Arab unity, and progressive Islam. Always the theorist rather than the practician, Qaddafi established a series of committees in the Basic People's Congress, which became somewhat disastrous after being led by zealots. The zealots were more concerned about persecuting violators than improving the state. It was illegal to protest in Libya and torture was a frequent means of keeping citizens in line with regime. The state was in charge of all activity, both political and economic and very little freedom existed. Anyone who left the country was hunted down and assassinated, and public executions broadcasted on television were common.[3]

Most problematic for the West was Qaddafi's sponsorship of terrorism throughout the world including Hamas, the Irish Republican Army and the Black Panthers. Qaddafi was suspected of involvement in blowing up an American plane over Scotland in 1988 and refused to turn over the perpetrators of the bombing until 1999. When he came into office, George W. Bush managed to temper the radical Qaddafi as the dictator allowed weapons inspectors into Libya and toned down his radical rhetoric, fearful of an American invasion. A few years later, Qaddafi's abusive and incompetent reign would soon come to an end under the presidency of Barack Obama.[4]

The conclusion of Qaddafi's life on October 20, 2011 and the end of his reign in Libya began with the Arab Spring. After Tunisia and Algeria, Libya was the third country to experience the Arab Spring in the Middle East.

Although many political and media pundits in the West believed that the Arab Spring occurred because Arabs suddenly wanted democracy after living under decades of corrupt dictatorships, this was not the case. Democracy was a nice concept for some people, particularly the wealthy, but the real reasons for the Arab Spring were swept under the rug by the same people who argued that the Arab Spring was the initial democratization revolution in the Middle East. How the West had hoped that the Arabs would finally be able to democratize and stop human rights abuses in their own countries!

Unfortunately, the lack of food due to a 60 percent increase in commodity prices is what caused the people throughout the Middle East to riot not dreams of fledgling democracies. This worldwide inflation was due to the American Federal Government's QE2 program, enacted by the Chair of the Federal Reserve, Ben Bernanke.[5] Three meals went to two meals a day and so forth; the standard of living decreased dramatically for the poor who were already struggling. Political protest became a vehicle for the now unemployed as daily living had become too difficult.

Libya was no stranger to this inflation. Libyans watched Muammar Gaddafi and his family live in luxury while they scrimped to get by; this decadent behavior was exhibited to the public in a series of Wikileaks taken from the US Embassy in Tripoli. Qaddafi even defended the Tunisian President Ben Ali after the initial revolution began in Tunisia. On February 15, 2011, protests broke out in Benghazi after the arrest of a human rights lawyer Faathy Terbil. Terbil had been arrested for serving as counsel for several families who lost their loved ones in a prison massacre of over 1200 people.[6] Qaffadi responded to protests by shooting several people and killing a dozen. Rebel forces in response broke into the military garrison and stole weapons to fight government forces. From there on, the violence continued to escalate.

In response, the United Nations Security Council (UNSC) adopted Resolution 1970 on February 26, 2011, which expressed "grave concern" over the situation in Libya and imposed an arms embargo on the country. Meanwhile the National Transitional Council (NTC) composed of Mustafa Abd al-Jalil, Fathi Tarbil and Abd al-Rahim, began claiming that they were the true representatives of the Libyan people. Many governments in the West listened to the academically composed NTC. The NTC fought to become legitimate in the eyes of the Libyan people.[7] Although the NTC managed to solidify their role, their lack of organization and communication later led to their downfall.

The initial Alliance deployed Airborne Warning and Control Systems (AWACS) aircraft to the area to provide round-the-clock observation. These "eyes-in-the-sky" gave NATO detailed information about movements in Libyan airspace. Two days later the Alliance moved ships from current NATO assets, as well as ships made available by NATO nations for the mission, to the Mediterranean Sea to boost the monitoring effort. A no-fly zone was declared over Libya in an effort to protect citizens from Qaffadi's retaliation after seeing gross human rights violations over satellites. On March 19, NATO began airstrikes against Qaffadi's forces, thus named Operation Odyssey Dawn. The Americans shot Tomahawk cruise missiles against Qaffadi's forces and the ports were blockaded against the influx of weapons and mercenaries. On August 23, 2011, the rebels, backed by NATO planes, overtook Qaffadi's compound, Bab al-Aziziya in Tripoli. Qaddafi had already fled to his hometown of Sirte by the time the rebels seized the compound.

The fighting lasted over eight months and Libyans realized that even though NATO had their backs, Qaffadi was harder to dislodge than imagined. The rebel forces were even assisted by the Islamic terrorist group al–Harakat al-Islamiyah al-Libiyah, who united with the people against Qaddafi.[8] In October 2011, a new interim transitional government was announced, complete with a draft constitution.

American intervention in Libya occurred for two reasons: The first was humanitarian in that Qaffadi would commit massive human rights abuses against his own people if the United States did not step in. The second argument revolved around American support for the Arab Spring. Since Libya had one of the earliest uprisings, Qaddafi's actions would set a precedent for the other dictators in the Middle East. If the United States allowed Qaddafi to slaughter his own people and refused to support what was viewed as Arab democratization, the Americans would be on the wrong side of the history. If President Obama took a stand, albeit insignificant against Qaddafi, the Americans could show solidarity and minimal support for the Libyans and other upcoming revolutions in the Middle East.[9] This would in turn bolster the reputation of the United States in the eyes of the Obama Administration in the Middle East and throughout the world.

Why Libya was President Obama's intervention of choice, concerning the Arab Spring is unclear. President Obama was elected on the premise that he would bring the troops home from Iraq and Afghanistan, and would no longer "put boots on the grounds" so putting troops in Libya was not an option. Obama threatened the president of Syria, Bashir al-Assad, several times over his blatant use of chemical and biological weapons against his own people, yet nothing was done about it. President Assad was still using chemical weapons against his own citizens in August, 2016. *Foundation for the Defense of Democracy* Fellow, Daveed Gartenstein-Ross adds that another reason for the United States to intervene in Libya was the al-Qa'ida adherents within the population.[10] Al-Qa'ida was strengthened by the downfall of Qaddafi as the collapse of the state, albeit an already weak one would provide the niche for al-Qa'ida to start recruiting and conquering territory within Libya. If the situation declined more, the country would be ripe for the jihadists to come in and establish Islamic Law.[11]

The next section of this chapter talks about drone warfare in Libya, establishing the numbers and statistics. The mistakes concerning the NATO intervention will also be discussed next. The chapter continues on to examine whether the Libyan intervention was successful or not. Lastly, in the conclusion, the problems with the Libyan intervention will be discussed and summarized along with specific problems that occurred during the Libyan intervention.

DRONE WARFARE AND TARGETED KILLING IN LIBYA

On March 19, 2011, the United States began attacking Libya under AFRICOM (United States Africa Command) (commanded by General Carter Ham) and the onslaught of advanced American weaponry and technology began to seep into Libya. As soon as NATO was ready to assume command of Operation Unified Protector, responsibility was quickly transferred to NATO on March 31, 2011 although still under the command of American General Carter Ham. The new name of the NATO forces became Operation Odyssey Dawn. General Ham was still under the U.S. European Command (EUCOM).[12]

Operation Odyssey Dawn struck some 6000 targets. Around 8000 personnel and twenty-one warships were included in the Operation.[13] NATO flew around 18,000 armed sorties, including France, the United Kingdom, and the United States. Warplanes from France, Britain, the United States, Italy, Norway, Denmark, Belgium, and Canada cooperated in the Operation. Two non-NATO nations, Qatar and the United Arab Emirates, also participated. After seven months, the NATO alliance had destroyed more than 5,900 military targets through 9,700 strike sorties. Approximately 7,600 Hellfire missiles were used. A small percentage of those armed sorties, approximately 250 were flown by drones and the Predator flew in 145 sorties.[14] "Strike sorties' is the term used for 'identifying and engaging targets', while armed sorties could also be for surveillance, and carrying weapons for self-defense."[15] According to *The Bureau of Investigative Journalism*, the Pentagon stated that the United States carried out 105 strikes between early April and September 2, 2011. However, the full campaign lasted until October 31, 2011. According to the numbers, more than one in three armed missions led to a drone strike but this figure does not reflect the full campaign, which continued until October 31, 2011.[16]

MISTAKES MADE BY NATO

Although there is no independent think tank that has begun to collect civilian casualty records in Libya, it is clear that mistakes were made and civilians were killed. Scattered news stories provide a picture of a NATO campaign that was not perfect. NATO and its allies stated that every effort was made to minimize civilian casualties. Following the end of the campaign, in November, 2011, NATO Secretary General Anders Fogh Rasmussen stated, "I have to say that we conducted our operations in Libya in a very careful manner, so we have no confirmed civilian casualties caused by NATO. And furthermore, we have really minimized the amount of collateral damage."[17]

However, *The New York Times* found that between 40 and 70 civilians were killed in the NATO bombings. Their investigation was based on: interviews with survivors, doctors, witnesses, collection of munitions remnants, medical reports, death certificates, and photographs. The investigation included site visits to more than twenty-five sites, including: Tripoli, Surman, Mizdah, Zlitan, Ga'a, Majer, Ajdabiya, Misurata, Surt, Brega and Sabratha, and near Benghazi. More than 150 targets were hit at these places.

One site visit found the remnants of NATO munitions. Problematically, NATO refuses to admit that it attacked the site. This mistake killed twelve anti-Qaddafi fighters, in addition to nearly killing an ambulance crew full of civilians who were helping the wounded. Similar to American drone strikes, the second NATO strikes led by the United States would return to kill civilians rushing to the wounded as many eye witnesses recounted.[18] Known as the "double tap" this technique is used to kill terrorist members who come to the strike area to claim bodies.

Another error occurred on June 19, 2011, when the family of Ali Mukhar al-Gharari was killed. An errant bomb hit his three story house he shared with his children. Four of his family members were killed and eight were wounded. NATO stated that it was trying to destroy a munitions compound although this was never validated.[19]

Another report compiled by *Human Rights Watch* found that "NATO air strikes killed at least 72 civilians, one-third of them children under age 18." The report was conducted from August 2011 to April 2012. NATO has rarely acknowledged most of these casualties or made any effort to examine how and why they occurred.[20] On June 19, 2011, a NATO bomb hit a family home in Tripoli, killing five civilians and wounding at least eight. In this instance, NATO admitted a mistake stating that "weapons system failure which may have caused a number of civilian casualties." However, no action was taken to correct this mistake or compensate the victims or their families.[21]

On August 29 or 30 in Bani Walid, a NATO strike hit two homes, killing five members of the Jfara family, including two men, two women, and a young girl. No evidence of military activity was found here either. NATO stated that it destroyed a "major command and control node which was reliant on non-traditional/informal methods to carry out that function." On September 25, 2011, NATO struck a home in Sirte, containing the brother of a senior military officer, Brig. Gen. Musbah Ahmed Dyab. Three women and four children from the family were killed. On August 4, 2011 in the town of Zliten, a NATO bomb hit the house of Mustafa al-Morabit. His wife and two children were killed. No sign of any military activity was found. Al-Morabit believed that NATO mistook his house for a neighbor's home, which Qaffadi forces had used and evacuated on August 2. NATO stated that it had hit a "command and control node" but evidence of this was not found.[22]

The most disastrous attack occurred in the village of Majer on August 8, 2011. Thirty-four people were killed and more than thirty were wounded when NATO bombed two family compounds. The second strike, which killed eighteen men hit a neighboring compound where people were assisting wounded and pulling the dead from the rubble. NATO stated that the compounds were a "staging base and military accommodation" for Qaffadi forces. Muammar al-Jarud, who was a victim of the attack stated, "I'm wondering why they did this, why just our houses. We'd accept it if we had tanks or military vehicles around, but we were completely civilians and you can't just hit civilians."[23]

The military aspect of the NATO intervention was successful as Qaffadi was found and executed by his own people. However, NATO did make some mistakes and less than 100 innocent people lost their lives. In the whole scheme of things, this loss of human life is not nearly as bad as some of the other countries in this book. However, the intervention was not perfect nor did it leave the people of Libya in a better situation.

DISCUSSION OF THE LIBYAN INTERVENTION

The data concerning drone strikes on the NATO Libyan intervention in which the United States led and participated is not extensive. This was an eight month campaign that used drone strikes and air strikes to assassinate Libyan forces and destroy Libyan compounds. Unlike other chapters in the book, there is not much to discuss concerning the exact effects of drone strikes on the Libyan people and country. However, the elephant in the room that should be addressed is whether Libya is better off after the NATO intervention where drone strikes were routinely used.

Unfortunately, the answer to that question is no. From a military standpoint, Operation Odyssey Dawn was effective as Qaddafi, the main objective was taken out of power. In addition, scholar Christopher Chivvis found that the Operation only cost the United States approximately 1.1 billion dollars. To put this number in perspective, the Kosovo air war in 2011 costed the United States about 2.7 billion dollars not including veterans' benefits.[24]

However, if success is to be gauged by the aftermath of Operation Odyssey Dawn, the Libyan intervention was not successful. Libya is definitely not in better shape since the dictator Qaddafi was taken out of power. In fact, Libya's situation is even direr now that the hated dictator is dead. The reasons for these deteriorating circumstances are numerous. The first reason is that Libya is now a failed state that has absolutely no power structure. After Qaffadi's death and transition to a new government with elections on July 7, 2012, Libya is still not doing well. Muhamad Magarief was elected the new President on August 8, 2012 and a new Constitution was drafted in 2013 but the Libyan Muslim Brothers won seventeen out of eighty seats in March 2012.[25] Pro-Qaddafi rebels use terrorism on a regular basis to create havoc. The American Consulate in Benghazi was attacked on September 11, 2012 by members from Ansar al-Sharia and four Americans were tortured and killed.[26] The NTC has created its own power vacuum and self-proclaimed ad-hoc police are roaming the streets providing vigilante justice. Clan and tribal identity is still a major obstacle for overcoming disunity.[27]

The second reason Libya is in worse shape after the revolution is the deteriorating economic situation. The economy was still in a recession in 2015. Oil production fell to 0.4 million barrels per day, which is approximately 25 percent of potential Libyan oil production. The GDP declined by 10 percent and the per capita income has fallen to less than 4,500 US dollars as compared to 13,000 US dollars in 2012. Inflation averaged 9.2 percent in 2015, driven by a 13.7 percent rise in food prices. The costs of flour and bread have quintupled. Overall, the budget deficit grew from 43 percent of GDP in 2014 to more than 75 percent of GDP in 2015.[28]

The third reason that Libya is in worse shape post-revolution is because Libya has become a new hotbed for ISIS as recruits are being sent to Libya on a daily basis to help extend the caliphate. In the past two years, at least two American airstrikes/drone strikes have targeted members of ISIS in Libya. Estimates of ISIS members in Libya now number over 5000.[29] In November 2015, strikes were used against Abu Nabil, an al-Qa'ida operative who had joined ISIL.[30] In mid-February, 2016, the United States conducted a series of strikes in Sabratha, Libya to target the Tunisian ISIS leader, Noureddine, Chouhane. The strikes were aimed toward a training camp that had been under surveillance for weeks. Forty-one people were killed in the strike, most of them Tunisians and some Algerians, although estimates of the dead range as high as sixty people.[31] ISIS was also responsible for several beheadings on an Egyptian Coptic Beach in early 2015 in Libya.[32]

The fourth reason that Libya is worse off after the NATO invention is that Libya is now a failed state and the country is ripe for al-Qa'ida and other jihadist training camps, etc,. to move in.[33] We already know that al-Qa'ida and ISIS have a strong presence in Libya. Without a strong central government, a legitimate police force, and a viable military, Libya will be fed to the wolves, the wolves being jihadist terrorist groups. The reasons for this are: it is easy to establish terrorist training camps, small arms running is a viable black market economy, terrorists use Libya as a safe haven, and Libya is the perfect place to plan future attacks.[34] While it is positive that a corrupt, possibly insane dictator like Qaddafi is no longer in power, there is no relief in sight for Libya.

To make matters worse, a few of Qaddafi's children are still alive. Ayesha Qaddafi, one of Muammar's daughters who also happens to be a lawyer swears to reestablish her father's republic. Ayesha lost her father, husband, and two children in the Arab Spring revolution. She is now distributing propaganda concerning the return of her family's rule as she vows to destroy NATO and Libya's terrorists.[35] Ayesha Qaddafi does have some support in Libya to reestablish her family's rule. A few of Qaddafi's sons are still alive but are most likely not interested in returning to Libya.

The fifth reason that Libya is worse off after American and NATO intervention is that the destabilization of Libya has led to greater problems for Libya's neighbors including Egypt, Tunisia, and Algeria. In addition, Mali suffered a jihadist takeover of Northern Mali that required an intervention by France. *Foundation for the Defense of Democracy* Fellow Gartenstein-Ross states, "It is not clear that lives were saved on the whole [concerning the Libyan NATO intervention]. As one of my contacts in the intelligence community recently put it, 'lives that were saved in Benghazi turned into lives lost in Timbuktu.'"[36] The spillover effect of Libya into other countries has managed to destabilize a whole area of Africa.

Unfortunately, it is almost nearly impossible to make the argument that Libya has improved after the NATO invention or that the lives of Libyans have improved. While it is a fact that an abominable dictator and his family were removed from power, Libyans face more danger on a daily basis concerning ISIS, rogue Islamist groups, and economic failure. Most likely the average person would have preferred living under an autocratic regime, fearful of criticizing Qaffadi rather than living in a chaotic political situation, rife with terrorism, and economic failure. A public opinion poll for the first ever *National Survey of Libya* conducted in 2012, found that the population would prefer a dictatorship over democracy. The survey which asked over 2,000 Libyans found that in five years' time 35 percent would still like a strong leader or leaders for the country. Only 29 percent would like to live in a democracy. Sixteen percent were ready to resort to violent means for political reasons if necessary.[37] Life under Muammar Qaddafi was predictable; daily life in Libya is not.

CONCLUSION

In conclusion, the intervention into Libya from a military standpoint was successful; Qaffadi was taken out of power. From a human rights perspective, there were a few problems. NATO has maintained in several cases, where *Human Rights Watch* and *Campaign for Innocent Victims in Conflict* investigated, that no mistakes were made and that civilians were not assassinated. Fred Abrahams, a researcher from *Human Rights Watch* stated that, "It's crystal clear that civilians died in NATO strikes. But this whole campaign is shrouded by an atmosphere of impunity."[38] The killing of Libyan civilians (led by the United States in NATO) is not a good American foreign policy. Very few, if any, mistakes should be made and reparations should be given to the victim's families if mistakes were made.

One of the major problems that occurred in Libya during the intervention, which is common for every country in this book, is the concept of the "double tap." In one instance on April 7, 2011, anti-Qaddafi forces outside Brega were destroyed by NATO forces, confusing them for Qaddafi forces. After the anti-Qaddafi's forces formations were destroyed, NATO came back for the "double tap" to kill survivors. Many of the survivors were civilians trying to help those that had been attacked including three shepherds in the area. The concept of the "double tap" needs to stop. There are several instances of this strategy in the book and it only seems to kill civilians. Perhaps initially in Afghanistan, the double tap was useful as al-Qa'ida or Taliban members would collect the bodies after a drone strike and would be easy targets. However, militant groups are now aware of this strategy and are no longer going to the scene of the drone strike.

NOTES

1. Muammar Qaffadi, *My Vision*, trans. Angela Parfitt (London: John Blake, ltd, 2005).
2. Ronald Bruce St. John, *Libya from Colony to Revolution*, (Oxford: One World, 2012).
3. Dirk Vandewalle, *A History of Modern Libya* (Cambridge: Cambridge University Press, 2006).
4. David S. Sorenson, *An Introduction to the Modern Middle East,* 2nd ed. (Boulder: Westview Press, 2014), Chapter 18.
5. Christine Sixta Rinehart, "Sixta Rinehart: Iran all Over Again," T*he State* (Columbia, SC), February 10, 2011, Editorial-opinion Extra.
6. Anjali Kamat and Ahmad Shokr, "Libya" in *Dispatches from the Arab Spring Understanding the New Middle East*, eds. Paul Amar, Vijay Prashad (Minneapolis: University of Minnesota Press, 2013).
7. Peter Bartu, "The Corridor of Uncertainty," in *The Libyan Revolution and Its Aftermath,* eds. Peter Cole and Brian McQuinn (Oxford: Oxford University Press, 2015).
8. Mary-Jane Deed, "The Arab Spring: Libya's Second Revolution," in *The Arab Spring Change and Resistance in the Middle East*, eds. Mark L. Hass and David W. Lesch (Boulder: Westview Press, 2013).

9. Christopher S Chivvis, *Toppling Qaddafi: Libya and the Limits of Liberal Intervention,* (Cambridge: Cambridge University Press, 2013).

10. Daveed Gartenstein-Ross, "Success and Failures of the U.S. and NATO Intervention in Libya," *Foundation for the Defense of Democracies,* May 1, 2014, accessed January 10, 2016, http://www.defenddemocracy.org/media-hit/gartenstein-ross-daveed-successes-and-failures-of-the-us-and-nato-intervention-in-libya/.

11. *Federal Research Division*, Library of Congress, "Al Qaeda in Libya: A Profile," August 2012, accessed April 25 https://fas.org/irp/world/para/aq-libya-loc.pdf.

12. Richard A. Lobban Jr. and Christopher H. Dalton, *Libya, History and Revolution* (Santa Barbara, CA: Praeger, 2014).

13. Christopher Chivvis, *Toppling Qaddafi, Libya and the Limits of Liberal Intervention*, (Cambridge: Cambridge University Press, 2014).

14. Chris Woods and Alice K. Ross, "Revealed: US and Britain launched 1,200 drone strikes in recent wars," *The Bureau Investigates*, December 4, 2012, accessed March 27, 2016, https://www.thebureauinvestigates.com/2012/12/04/revealed-us-and-britain-launched-1200-drone-strikes-in-recent-wars/.

15. Chris Woods and Alice K. Ross, "Revealed: US and Britain launched 1,200 drone strikes in recent wars," *The Bureau Investigates*, December 4, 2012, accessed March 27, 2016, https://www.thebureauinvestigates.com/2012/12/04/revealed-us-and-britain-launched-1200-drone-strikes-in-recent-wars/.

16. Chris Woods and Alice K. Ross, "Revealed: US and Britain launched 1,200 drone strikes in recent wars," *The Bureau Investigates*, December 4, 2012, accessed March 27, 2016, https://www.thebureauinvestigates.com/2012/12/04/revealed-us-and-britain-launched-1200-drone-strikes-in-recent-wars/.

17. Anders Fogh Rasmussen, "Monthly Press Briefing," *North Atlantic Treaty Organization*, November 3, 2011, accessed April 22, 2016, http://www.nato.int/cps/en/natolive/opinions_80247.htm.

18. C.J. Chivers and Eric Schmitt, "In Strikes on Libya by NATO, an Unspoken Civilian Toll," *The New York Times*, December 17, 2011, accessed April 22, 2016, http://www.nytimes.com/2011/12/18/world/africa/scores-of-unintended-casualties-in-nato-war-in-libya.html?pagewanted=all&_r=2&.

19. C.J. Chivers and Eric Schmitt, "In Strikes on Libya by NATO, an Unspoken Civilian Toll," *The New York Times*, December 17, 2011, accessed April 22, 2016, http://www.nytimes.com/2011/12/18/world/africa/scores-of-unintended-casualties-in-nato-war-in-libya.html?pagewanted=all&_r=2&.

20. Fred Abrahams and Sidney Kwiram, "Unacknowledged Deaths," *Human Rights Watch*, May 13, 2012, accessed May 3, 2016, https://www.hrw.org/report/2012/05/13/unacknowledged-deaths/civilian-casualties-natos-air-campaign-libya.

21. Fred Abrahams and Sidney Kwiram, "Unacknowledged Deaths," *Human Rights Watch*, May 13, 2012, accessed May 3, 2016, https://www.hrw.org/report/2012/05/13/unacknowledged-deaths/civilian-casualties-natos-air-campaign-libya.

22. Fred Abrahams and Sidney Kwiram, "Unacknowledged Deaths," *Human Rights Watch*, May 13, 2012, accessed May 3, 2016, https://www.hrw.org/report/2012/05/13/unacknowledged-deaths/civilian-casualties-natos-air-campaign-libya.

23. Fred Abrahams and Sidney Kwiram, "Unacknowledged Deaths," *Human Rights Watch*, May 13, 2012, accessed May 3, 2016, https://www.hrw.org/report/2012/05/13/unacknowledged-deaths/civilian-casualties-natos-air-campaign-libya.

24. Christopher Chivvis, *Toppling Qaddafi, Libya and the Limits of Liberal Intervention*, (Cambridge: Cambridge University Press, 2014), 176–177.

25. Richard A. Lobban Jr. and Christopher H. Dalton, *Libya, History and Revolution* (Santa Barbara, CA: Praeger, 2014).

26. Simon Chase and Ralph Pezzullo, *Zero Footprint* (New York: Mulholland Books, 2016).

27. Fahed al-Sumait, Nele Lenze, and Michael C. Hudson, *The Arab Uprisings, Catalysts, Dynamics, and Trajectories* (Lanham: Rowman & Littlefield, 2015).

28. *The World Bank*, "Overview Libya," March 31, 2016, accessed May 5, 2016, http://www.worldbank.org/en/country/libya/overview.

29. Al Jazeera and Agencies, "ISIL training camp' in Libya targeted by US," *al-Jazeera*, February 20, 2016, accessed April 20, 2016, http://www.aljazeera.com/news/2016/02/deaths-reported-raids-isil-camp-libya-160219131122223.html.

30. BBC News, "Libya IS head 'killed in US air strike'," *BBC News*, November 15, 2015, accessed April 20, 2016, http://www.bbc.com/news/world-africa-34823466.

31. Al Jazeera and Agencies, "ISIL training camp' in Libya targeted by US," *al-Jazeera*, February 20, 2016, accessed April 20, 2016, http://www.aljazeera.com/news/2016/02/deaths-reported-raids-isil-camp-libya-160219131122223.html.

32. Greg Botelho and Barbara Starr, "49 killed in U.S. airstrike targeting terrorists in Libya," *CNN*, February 20, 2016, accessed April 25, 2016, http://www.cnn.com/2016/02/20/africa/libya-us-airstrike-isis/.

33. *Federal Research Division*, Library of Congress, "Al Qaeda in Libya: A Profile," August 2012, accessed April 25, https://fas.org/irp/world/para/aq-libya-loc.pdf.

34. Daveed Gartenstein-Ross, "Success and Failures of the U.S. and NATO Intervention in Libya," *Foundation for the Defense of Democracies,* May 1, 2014, accessed January 10, 2016, http://www.defenddemocracy.org/media-hit/gartenstein-ross-daveed-successes-and-failures-of-the-us-and-nato-intervention-in-libya/.

35. Editor's Choice Libya, "Ayesha Qaffadi as a new leader of resistance against NATO and the Libyan terrorists," *South Front*, January 2, 2016, accessed April 28, 2016, https://southfront.org/ayesha-Qaffadi-as-a-new-leader-of-resistance-against-nato-and-the-libyan-terrorists/.

36. Daveed Gartenstein-Ross, "Success and Failures of the U.S. and NATO Intervention in Libya," *Foundation for the Defense of Democracies,* May 1, 2014, accessed January 10, 2016, http://www.defenddemocracy.org/media-hit/gartenstein-ross-daveed-successes-and-failures-of-the-us-and-nato-intervention-in-libya/.

37. *University of Oxford*, "University of Oxford: National Public Opinion Poll of Libya," February 14, 2012, accessed May 30, 2016, http://www.prnewswire.com/news-releases/university-of-oxford--national-public-opinion-poll-of-libya-139308408.html.

38. C.J. Chivers and Eric Schmitt, "In Strikes on Libya by NATO, an Unspoken Civilian Toll," *The New York Times*, December 17, 2011, accessed April 22, 2016, http://www.nytimes.com/2011/12/18/world/africa/scores-of-unintended-casualties-in-nato-war-in-libya.html?pagewanted=all&_r=2&.

Chapter 6

Iraq and Syria

The Genesis of ISIS

THE HISTORY OF ISIS

In its original conception, this book was supposed to have separate chapters on Syria and Iraq. However, when looking at the data on targeted killing in both Iraq and Syria, it occurred to the author that most of the targeted killing occurred for the purposes of destroying some component of ISIS. Since ISIS now occupies parts of both Syria and Iraq, it was decided that it was easiest to combine both Iraq and Syria into one chapter since much of the history of the group concerns both countries. ISI came out of AQI or al-Qa'ida in Iraq and this group later announced its affiliation with al-Qa'ida in Syria in May 2013. After the two branches united, the entire Iraqi and Syrian group became known as ISIS or ISIL, depending on one's favored term and perspective. ISIS refers to the Islamic State in Iraq and Syria. ISIL refers to the Islamic State of Iraq and the Levant. The Levant is defined as the lands around Syria or "lands of the rising sun," including Jordan, Israel, Palestine, and Syria.

As an introductory precaution, this chapter offers more of a description of the effects of targeted killing on ISIS in Iraq and Syria. Since target killing began in Iraq and Syria in 2014, there is not much data or research to analyze or write about at this point. With only two years of data to work with, there cannot be any solid conclusions made at this time about the effects of targeted killing on ISIS in Iraq and Syria.

With that being said, we shall start at the beginning to tell the story of ISIS in Iraq and Syria. On September 11, 2001, the United States was attacked by nineteen al-Qa'ida members who hijacked four American planes. The suicide missions were successful when two of the planes flew into the World Trade Center in New York City, eventually bringing the towers to the ground. The third plane hit the Pentagon outside of Washington D.C. and the fourth plane crashed into a field in Pennsylvania after passengers wrestled the attackers to the floor. Around 3000 people died that day and the acts of American heroism were astounding. Rallying around the flag, the American people supported George W. Bush when he invaded Afghanistan on October 7, 2001 to bring down al-Qa'ida and the Taliban, the perpetrators and sponsors of 9–11.[1]

With the thoughts of 9–11 still emblazoned in the memories of most Americans and the current war in Afghanistan, it was purported that America's and Bush's long-time foe Saddam Hussein had been providing aid and assistance to Osama bin Laden, the mastermind of 9–11. President Hussein was also cited as a state sponsor of terrorism, particularly for Palestinian suicide bombers. In addition, it was found that Saddam Hussein was creating weapons of mass destruction, particularly nuclear weapons and could possibly threaten Israel. Lastly, Bush declared Hussein's flagrant abuse of human rights as his fourth reason to invade Iraq. After publicly asking Saddam Hussein to stop pursuing nuclear weapons and asking the United Nations' assistance for help on several occasions, Bush finally gave up and attacked his father's foe.

On March 20, 2003, President George W. Bush began a series of military strikes on Iraq in an attempt to disable the Iraqi military and assassinate the brutal Iraqi dictator, Saddam Hussein. The War was quickly won from a military standpoint and some Iraqis assailed the streets, welcoming their American liberators and toppling Saddam Hussein statues and strongholds. Saddam's Baathist Party elites seem to disappear into the woodwork in Iraq, which would cause problems later. President Bush was seen atop the American Air Force carrier *Abraham Lincoln* declaring "Mission Accomplished" although in reality, the mission was going to be much more difficult to accomplish than anyone had ever thought.[2]

After the initial invasion of Iraq, what followed was a ground war for hearts and minds of the Iraqi people. As Iraqi rebels strung up the bodies of Blackwater private security guards who worked for the United States, Iraqis voted in their first election on January 30, 2005.[3] American military forces began taking towns and cities from the rebels piece by piece. The Abu Ghraib Prison scandal managed to put a dent in the American ideals of a "just war" as pictures were shown on national television of American servicemen and women abusing and humiliating Iraqi prisoners. After years of fighting insurgents, roadside bombs, and IEDs, the United States finally left Iraq on December 15, 2011. Over 4400 American military men and women had died and the United States had spent over 1.7 trillion with an additional 490 billion in benefits that will be paid to war veterans over the next several decades. The total cost of the Iraq War came to around 2.19 trillion with included estimated veteran benefits, although the reasons for invading Iraq are still hotly debated today.[4,5]

After American forces evacuated, a power vacuum was created in Iraq. There was no powerful stable military force to keep order in a chaotic country. Kurds, Sunnis, and Shi'ites continued to vie for power in a free-for-all, failed state of Iraq. A major future player in the Iraqi saga, Abu Mus'ab al-Zarqawi (Ayman al-Zarqawi) was released from al-Sawwaqa prison in Jordan directly before joining al-Qa'ida in Afghanistan in 1999. Al-Zarqawi eventually fled Afghanistan (briefly stopping in Iran) for Iraq after the Taliban fell somewhere in 2001 or 2002. Zarqawi believed that the United States would eventually invade Iraq and this was his chosen location for his jihad. He did not have to wait too long as the United States invaded in March 2003.

Zarqawi pledged his oath of allegiance to the leaders of al-Qa'ida, Osama bin Laden and Ayman al-Zawahiri in October 2004, creating AQI. Only containing 10 percent Iraqis, the rest of AQI consisted of foreigners. Zarqawi despised Shia Muslims, refusing to include them in AQI, persecuting them regularly in Iraq. However, Zarqawi was too violent even for al-Qa'ida as Zawahiri criticized Zarqawi for attacking Shia Muslims in Iraq in an effort to start a sectarian civil war.[6] Zarqawi routinely targeted Muslims in suicide bombings and al-Qa'ida was not entirely comfortable with purposefully targeting the local Muslim civilians. However, problems with Zarqawi were ended when an American airstrike killed him on June 7, 2006. Shortly after his death, Zarqawi's men, led by Abu Ayyub al-Masri declared the creation of the Islamic State in Iraq much to the surprise and chagrin of Osama bin Laden and Ayman al-Zawahiri. Bin Laden and Zawahiri had wanted Zarqawi to declare an Islamic State complete with the establishment of the caliphate after the American forces withdrew from Iraq, not beforehand.[7]

Now that the caliphate was established with the caliph Abu Umar al-Baghdadi (a shadowy figure that may have never existed) at its head, ISI tried to repair relations with al-Qa'ida. ISI declared their oath of allegiance to al-Qa'ida and relations appeared to improve. However, Abu Ayyub al-Masri, the actual leader of ISI was intent upon creating his own Islamic state versus another terrorist organization like al-Qa'ida. Al-Qa'ida did not give any approval for the declaration of an Islamic state nor did it like the primary allegiance of all ISI members to the Islamic state not al-Qa'ida. However, a united front of al-Qa'ida and ISI was publicly presented.[8]

As time progressed, ISI became somewhat of a failure. The original bounty placed on al-Masri's head by the United States was five million in 2007 but by May 2008, the bounty had fallen to 100,000 dollars. Some of this fall from grace was due to Masri's inability as a manager and commander. Masri believed in the imminent apocalypse, ordering his followers to build pulpits for the *Mahdi* (end of the world redeemer of Islam) to ascend into the major Muslim mosques like that in Medina.[9] He also ordered his followers to take over Iraq only to call them back less than a week later as they were spread too thinly across Iraq. Masri appeared a bit off kilter, making many jihadists and intelligence officials question his sanity. To add to it, Masri picked another caliph, Hamid al-Zawi, a former police officer who had joined ISI.

Masri's arbitrary choice of a new caliph would normally require the debate and affirmation of religious scholars. Similar to choosing the Pope, the designation of the caliph should be a long arduous process complete with divine study and worship. When these problems concerning the caliphate were pointed out, Masri killed dissenters, including members of the Islamic army in Iraq who had refused to swear allegiance to ISI. Shia Muslims were also killed for their faith as Masri believed they were infidels.[10]

ISI members came from all sectors of society and were not properly vetted when they swore allegiance to ISI. Members of Saddam Hussein's guard and anyone who wanted to join the state were admitted. After all, this was a new country not an Islamic organization. Masri, remote from the battlefield and sequestered, rarely knew what was going on. Numerous complaints from other Islamic organizations were sent to Masri as members of other terrorist organizations such as Ansar al-Sunna were tortured to death by Masri's men. These complaints fell on deaf ears, whether due to rebels within ISI or the incompetence of Masri, the reasons are unknown. Eventually groups like Ansar al-Sunna began talking to American troops about getting rid of ISI in late 2006. It got so bad that even Osama bin Laden started receiving complaints about ISI, who had initially pledged loyalty to al-Qa'ida. Bin Laden refused to chastise ISI and eventually both al-Masri and Abu Umar al-Baghdadi were killed by an American and Iraqi joint raid on Masri's compound in April 2010. Abu Bakr al-Baghdadi became the next caliphate for ISI, and was also unknown to his followers and lacked the appropriate religious pedigree.[11]

By the time American troops departed in December 2011, ISI has been discredited and disowned by al-Qa'ida but would rise again with the oncoming civil war in Syria. Syrian president Bashir al-Assad had been funneling jihadists fighters into Iraq since American troops invaded. One study even suggested that as many as 90 percent of Iraqi ISI members had come through or from Syria.[12] When Syrians began protesting the al-Assad regime in 2011, Assad in retaliation let several jihadists out of prison, which in fact led to the ISIS creation in Syria and their rebirth in Iraq.

It was actually Ayman al-Zawahiri that ordered the Iraqi Islamic State (ISI) to "form a group and send it to [Syria]"[13] and Abu Muhammad al-Jawlani was sent to oversee the effort in Syria. Originally calling itself the Nusra Front (Jabhat al-Nusra), the group only had about 200 members in August 2012 and at first, was as abusive to the local population as ISI was in Iraq. Al-Jawlani then expanded into the North but for some reason stopped performing suicide attacks against the locals. Scholar of militant Islam, William McCants contributes this to the presence of Abu Khalid al-Suri, a propagandist and philosopher in Syria who advocated a "hearts and minds" campaign toward the Syrian population. At this point, ISI in Iraq and the Nusra Front in Syria were at odds concerning strategy and what to do with stolen Syrian oil reserves. In addition, al-Baghdadi the caliphate of ISI was frustrated with al-Nusra as the men in Syria were more loyal to their leader, al-Jawlani than him. Pulling a power play, al-Baghdadi officially announced that al-Nusra was officially a part of ISI on April 9, 2013. In retaliation, al-Jawlani declared al-Nusra's independence from ISI. Baghdadi then appealed to al-Zawahiri (the leader of al-Qa'ida) threatening a schism if al-Nusra did not comply, as ISI was still officially under al-Qa'ida rule. Zawahiri publicly announced that both ISI and al-Nusra were under al-Qa'ida and ordered al-Suri to mediate between the two parties. Al-Suri quickly became disgusted with ISI's conduct and could not come to any agreement with either party. Frustrated with the attitude of both parties, Zawahiri then disaffiliated ISI and al-Nusra from al-Qa'ida in February 2013, stating that they were now both independent groups.[14]

The conflict then exploded into a social media battleground with various jihadists and religious scholars tweeting and posting their own thoughts on the squabble. ISI, in retaliation for criticism for killing and abusing Muslims, sent three assassins to kill Abu Khalid al-Suri. They were successful and ISI continued to steamroll their own path of conquering Iraq. Many jihadists continued to flock to Iraq to join the winners of the dispute or perhaps what was viewed as the new Caliphate in the Middle East. In March 2013, ISI moved into Raqqa, Syria, taking over the provincial capital. It continued to take cities and towns throughout Syria, and thus the name ISIS was attributed to the group, the Islamic State in Iraq and Syria.[15] Meanwhile, al-Nusra returned to al-Qa'ida's bosom and continues to fight against Syrian forces (under Bashir al-Assad) and American intervention in Syria. Al-Nusra currently presents itself as a softer alternative to ISIS as it tries to work with the civilian populations, instead of beheading them.

In June 2014, ISIS moved into Mosul, eventually increasing its territory to the outskirts of Aleppo, Syria. With this, the ISIS caliphate Abu Bakr al-Baghdadi was renamed Caliph Ibrahim al-Baghdadi on June 29, 2014 with pomp and circumstance, claiming its authority over all Muslims. The dictatorial president of Syria, Bashir al-Assad has been somewhat content to let ISIS exist within his borders in the midst of a civil war. Bashir was sure that Syrians would prefer his brutal regime over the vulgar, bloodthirsty apostates who had taken over Iraq. In addition, crying "terrorism" was also one way to get foreign assistance and save his regime.[16] On November 10, 2014, ISIS announced that it had received oaths of allegiance from jihadists (including Boko Haram) in Egypt, Libya, Yemen, Algeria, and Saudi Arabia although some of these groups that announced their support cannot be taken seriously.

Indeed, any jihadists who becomes disenchanted with their commander is able to run off and join the Islamic state of ISIS.[17]

In 2014–2015, it was estimated that ISIS had between 20,000 to 30,000 members with its peak estimated at 33,000. However, recent estimates in 2016 put ISIS at 19,000 to 25,000 members and this number has remained stable.[18] Whether this is due to the fact that several thousand ISIS members were killed by American airstrikes in Kobani with losses in Bayli and Ramadi in 2015 or to desertion, is unknown. Many ISIS members are leaving and heading to Libya in an effort to expand ISIS as Libya now has around 5000 ISIS members within its borders.[19] As will be discussed in the following paragraphs, ISIS numbers are actually unknown and vary according to expert.

In comparison, ISIS is more successful than al-Qa'ida as it functions as a state, insurgent force, terrorist organization, and light infantry force. One estimate states that ISIS was making as much as two million per day in September 2014 from gas, oil, farming, taxes, extortion, kidnapping for ransom, and illegal antique sales in addition to many other trades.[20] The more territory that ISIS conquers, the more income they steal. New recruits receive military training but not as much religious training like al-Qa'ida. Military training consists of pistols, assault rifles, rocket-propelled grenades and at times, mortars. New recruits are initially placed on guard duty and then are sent to the front line (including terrorist operations) if they prove themselves trustworthy on guard duty.[21]

It is not completely clear who is to blame for the creation and uprising of ISIS. One could look to George W. Bush for invading Iraq, dismantling the strongman regime of Saddam Hussein who would have crushed ISIS, had it even been allowed to begin in the first place. Or one could also look to President Obama for bringing American troops home from Iraq and ignoring the civil war in Syria, while simultaneously offering empty threats to Bashir al-Assad if he used chemical weapons on his own citizens. Of course, al-Assad did use weapons on his own people on numerous occasions and Obama did nothing. After all, President Obama has mostly ignored ISIS from 2012 to 2014, which has allowed it to proliferate.

More pointedly and perhaps another perspective is that it is not the job of the United States to police the world, most importantly because Muslims do not want America to solve their problems. Moderate Muslims must be held responsible for their acquiescence in the rise of the terrorist group ISIS. If there are indeed moderate Muslims, then they must stand up. In the end, it is the job of the citizens to fight for the freedom and integrity of their government and state, not the Americans. The next section will look at the descriptive effects of targeted killing on ISIS.

THE EFFECTS OF TARGETED KILLING ON ISIS

Drone Strikes in Iraq and Syria-Operation Inherent Resolve

In Table 6.1: *American Drone Strikes during the Iraq War*, the number of American drones strikes during the Iraq War can be seen.[22] There is no published data available before 2008, although we know that drone strikes were being used in Yemen in 2002, so the American military and CIA did have access to drones. In Chapter 4 of Chris Woods' book *Sudden Justice,* it is evident that drones were used a lot during the Iraq War before 2008. Retired sensor operator Michael Haas who fought during the Iraq War stated that,

> I think it [drone use] started becoming a preferred weapon. We were shooting so damned many of them we were actually running out of missiles at one point. They weren't expecting JTACs [forward air controllers] to use it readily. But they liked it because it could take out a building, it could take out a moving care, and there wasn't a big potential to frag any innocents. I think they went a little overboard with it.[23]

Most of the sorties in Iraq listed in the table below were done for the purposes of surveillance but as is visible in the table, drone strikes were only used approximately forty-eight times from 2008 to 2012. President Obama did not want to use any drone strikes in Iraq trying to fulfill domestic political campaign promises and ordered only five drone strikes between 2009 and 2011. When the United States left Iraq in December 2011, the strikes ceased for a few years until ISIS arrived.

The drone strikes in Iraq began again on August 7, 2014 (even though the drones had been performing surveillance operations since June, 2014) as the United States began targeting ISIS in Iraq. These drone strikes were both used for targeted killing and surgical strikes in Iraq to destroy targets like ISIS mortar positions. President Obama began these "limited strikes" but refused

Table 6.1 American Drone Strikes during the Iraq War

Iraq	*Total Armed Drone Sorties*	*Total Missiles Fired by Drones*	*Total Drone Strikes*
2008	5558	60	43
2009	5300	4	4
2010	3378	0	0
2011	2773	1	1
2012	0	0	0

Source: Chris Woods and Alice K. Ross, "Revealed: US and Britain launched 1,200 drone strikes in recent wars," The Bureau Investigates, December 4, 2012, accessed March 27, 2016, https://www.thebureauinvestigates.com/2012/12/04/revealed-us-and-britain-launched-1200-drone-strikes-in-recent-wars/.

to "allow the United States to be dragged into fighting another war in Iraq." This campaign began known as Operation Inherent Resolve initially in Iraq.

A month later on September 11, 2014, these strikes were expanded as President Obama announced his intent to destroy ISIS elements in Syria and Iraq. President Obama stated,

> America will lead a broad coalition to roll back this terrorist threat. Our objective is clear: We will degrade, and ultimately destroy, ISIL through a comprehensive and sustained counterterrorism strategy.
>
> First, we will conduct a systematic campaign of airstrikes against these terrorists. Working with the Iraqi government, we will expand our efforts beyond protecting our own people and humanitarian missions, so that we're hitting ISIL targets as Iraqi forces go on offense. Moreover, I have made it clear that we will hunt down terrorists who threaten our country, wherever they are. That means I will not hesitate to take action against ISIL in Syria, as well as Iraq . . .
>
> In June, I deployed several hundred American service members to Iraq to assess how we can best support Iraqi security forces. Now that those teams have completed their work—and Iraq has formed a government—we will send an additional 475 service members to Iraq. As I have said before, these American forces will not have a combat mission—we will not get dragged into another ground war in Iraq.[24]

In Iraq, ISIS operations concentrated on Sunni urban regions and transport routes through those regions, including the Anbar and Ninawa provinces. In Syria, ISIS attacks focused on the northeast and east regions of the country including border grounds with Turkey and Iraq. ISIS attacks in Syria primarily target opposition groups so that ISIS can establish its power and territory.[25]

As a result of Operation Inherent Resolve as can be seen in Figure 6.1: *American Drone Strikes in Iraq, 2014–2015*,[26] drone strikes have increased tremendously over the past two years in Iraq as Obama has tried to fight ISIS without putting any boots on the ground. Drone strikes are currently reaching more than 500 strikes per month in Iraq. As of March 29, 2016, the U.S. and coalition have conducted a total of 11,230 strikes with 7,556 in Iraq. The countries that have participated in the strikes in Iraq include: Australia, Belgium, Canada, Denmark, France, Jordan, The Netherlands, and the United Kingdom.[27] In some cases, the United States is assisting the Kurds in fighting ISIS by providing air support for Kurdish forces involved in the fight against ISIS. The Kurds have asked both NATO and President Obama to protect Kurdish forces with air power.

As can been seen in Figure 6.2: *American Drone Strikes in Syria, 2014–2015*,[28] drone strikes in Syria were relatively stable, around 200 to 300 strikes per month since it began in September 2014. However in December 2015 strikes skyrocketed to over 3000 in Syria and these high numbers have

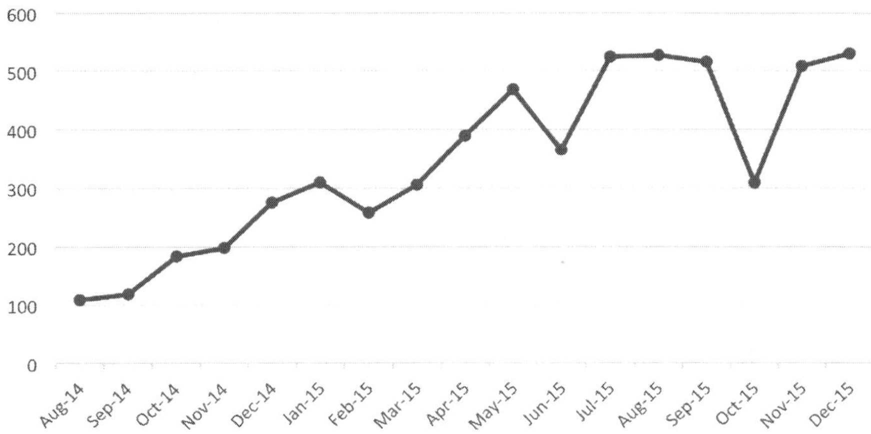

Figure 6.1 American Drone Strikes in Iraq, 2014–2015. *Source*: Chris Woods, "Strikes in Iraq," Airwars.org, 2016, accessed January 1, 2016, http://airwars.org/.

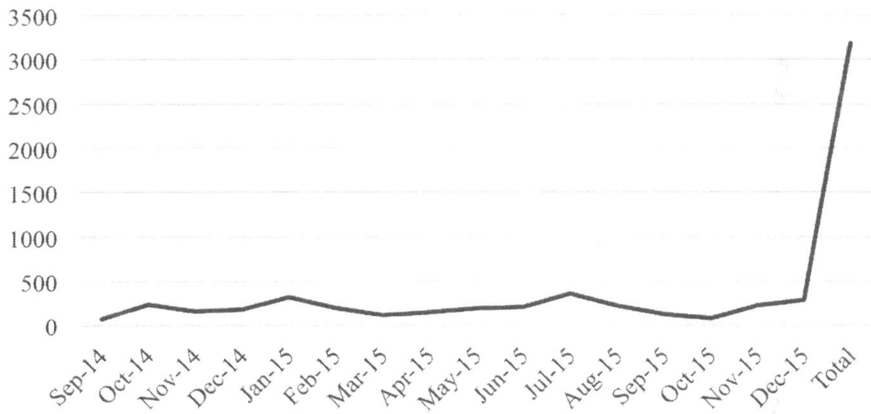

Figure 6.2 American Drone Strikes in Syria, 2014–2015. *Source*: Chris Woods, "Strikes in Syria," Airwars.org, 2016, accessed January 1, 2016, http://airwars.org/.

continued into 2016. These strikes skyrocketed because ISIS began expanding operations in Libya and Pakistan and expanded attack areas.[29] In addition, leaked documents from ISIS were published concerning the creation of an Islamic state throughout the world.[30] When the actual plans of ISIS were published, their intentions of creating an Islamic state through the world were finally taken seriously and President Obama realized he was no longer dealing with the "JV basketball team." As of March 29, 2016 the U.S. and coalition forces have conducted a total of 11,230 strikes with 3,674 in Syria. In Syria, the following countries have participated with the United States in the drone strikes: Australia, Bahrain, Canada, France, Jordan, The Netherlands, Saudi Arabia, Turkey, the UAE, and the United Kingdom.[31]

As a side note, in Iraq and Syria, there is also the problem that ISIS is committing genocide or ethnic cleansing against the Yazidi people and other people that are captured. The Yazidis are of Kurdish decent and are a Christian sect of pre-Islamic origin. A combination of Christianity, Judaism, and Zoroastrianism, the Yazidi religion is looked at as a polytheist religion by ISIS[32] and therefore ISIS has used their Muslim clerics to justify the enslavement and selling of Yazidi women and children. Women and young girls are used as sex slaves in the ISIS camps and have even been sold in slave markets after ISIS captured Mosul.

What Do We Know about Drone Casualties in Iraq and Syria?

One observation that can be made is that the number of casualties in Iraq, whether due to ISIS, American air strikes or drone strikes, a sectarian civil war, or all three has increased tremendously. Data on Iraqi deaths due to drone strikes are simply not available as no one is investigating drone strike incidents in Iraq. *Airwars* has found that the casualty rate from coalition forces is somewhere between 578 to 732 deaths from August 8, 2014 to June 30, 2015.[33] As can be seen in Figure 6.3: *Deaths in Iraq by Month, 2003–September 2015*,[34] the number of casualties in Iraq has increased to several thousand per month in recent months with as many as 4000 deaths in July 2014. The period after the United States evacuated Iraq was one of the most peaceful time periods concerning casualties that the country has had since before the United States invaded Iraq in March 2003. Since summer 2013, Iraqi deaths have increased dramatically.

There have been some major mistakes made in Iraq by the American forces. One such mistake occurred in al Hatra on March 13, 2015 when two Warthogs, also known as A-10's accidently fired on two vehicles at a check point outside the city. The city was controlled by ISIS. The Air Force claimed

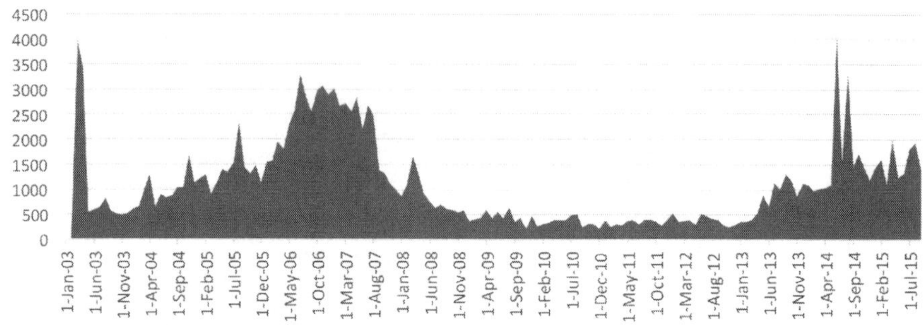

Figure 6.3 Deaths in Iraq by Month, 2003–September 2015. *Source*: Iraq Body Count, 2016, accessed March 27, 2016, https://www.iraqbodycount.org/.

responsibility for this attack that killed two women and three children who were cited as noncombatants. The pilots simply could not see the people they were firing on.[35]

In Syria, the *Syrian Network for Human Rights*, found that as of September 23, 2014 until February 13, 2016, the International Coalition has killed 267 civilians including ninety-three children and fifty women.[36] This coalition is led by the United States with the assistance of other nations including Australia, Bahrain, Canada, France, Jordan, The Netherlands, Saudi Arabia, Turkey, the UAE and the United Kingdom. Russian attacks that purposely targeted civilians in ISIS areas in Syria have led to the deaths of 15,027 civilians alone since September 23, 2014.[37] From March 2011 until October 31, 2015, 186,102 people have died in Syria as a result of the Civil War, Armed Opposition Groups, ISIS, Russian Forces, and International Coalition Forces as can be seen in Figure 6.4: *Syrian Death Toll, March 2011–October 31, 2015.*[38] It is evident that Syrian Government forces loyal to Bashir al-Assad have committed a majority of these murders in Syria. In the whole scheme of things, American drone strikes are not responsible for that many deaths although the amount of Syrian deaths from American drone strikes are likely to increase astronomically as the United States will continue its increase in drone strikes in 2016.

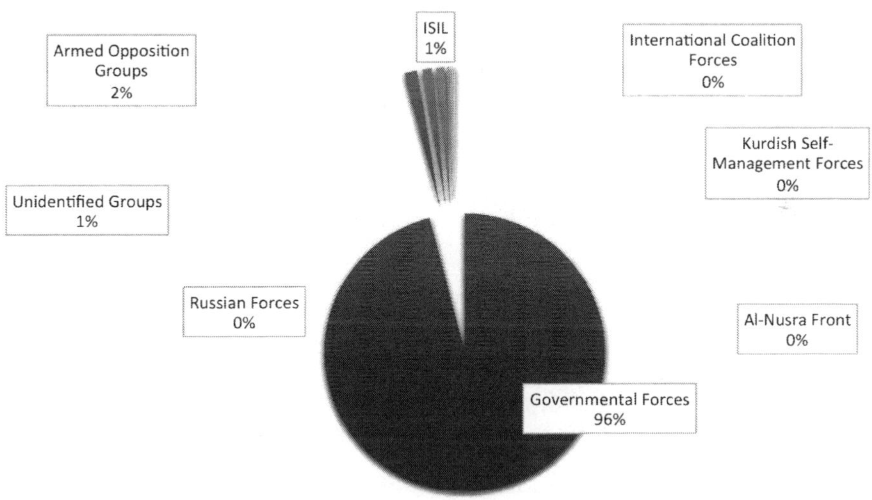

Figure 6.4 Syrian Death Toll, March 2011–October 31, 2015. *Source*: Syrian Network for Human Rights, "The Main Conflict Parties Who are Killing Civilians in Syria," Syrian Network for Human Rights, November 19, 2015, accessed April 12, 2016, http://sn4hr.org/wp-content/pdf/english/68_civilians_killed_in_new_attacks_by_international_coalition_forces_en.pdf.

Although data on the deaths of Syrians from American drone strikes is not available in total, there is still evidence of mistakes that have been made. It is evident that the United States and coalition forces are killing more civilians than they are acknowledging, which is typical in warfare situations for any country at war. Al Gharra was a mud-brick village in the Syrian Desert that was populated by descents of Bedouin tribes. The town has been held by: the Syrian Government, the Free Syrian army, ISIS, and now the Kurds. The United States used drones to attack ISIS elements that had invaded Al Gharra, offering support to the Kurds on the ground. This led to the destruction of the town. Abdul-Aziz al Hassan was the eldest of 11 children. His father who was the provider of ten children was killed in a drone strike on May 6, 2015 and the family's house was destroyed. Al Hassan, now a refugee in Turkey with his wife, kids, and siblings, asks the United States, "Even if we went back [to al Gharra, Syria], where would we live? In our destroyed house? Does the American government think we have money? Do they think that I can just go back and rebuild our house?"[39] Problematically the United States has never accepted any responsibility for this attack and most importantly ISIS elements were not present when the strikes occurred.

On December 7, 2015, a similar mistake was made in al Khan, Syria where forty-seven civilians were killed in an effort to kill ISIS members. Abu Khalil was awoken by a bomb at night. His mother and aunt were killed after the explosion when an Apache Helicopter came back to kill the leftovers. Abu stated, "It was the Americans. For the past year-and-a-half, the only aircraft that fly over our area have been American."[40] As has been reiterated in this book, when mistakes are made and entire villages and families are destroyed by American drone strikes, the possibility of empathy and recruitment for terrorist organizations is more likely.

According to public opinion polls in September 2015, Iraqis and Syrians think that their countries are heading in the wrong directions. Sixty-six percent of those questioned in Iraq and 57 percent in Syria think their countries are heading in the wrong directions. The problematic thing about this poll is that it also found that 25 percent of Iraqis thought their country was heading in the right direction and 33 percent of Syrians felt that Syria was heading in the right direction. Fifty-six percent of Iraqis oppose coalition air strikes while 50 percent of Syrians oppose coalition air strikes.[41]

The Effects of Drone Strikes on ISIS Numbers and Equipment

We are not really sure about ISIS's numbers and the estimates vary drastically depending on the source. In August, 2014, security expert Hisham al-Hashimi claimed that ISIL's total membership could be closer to 100,000.[42] Again in August 2014, Rami Abdel Rahman, Director of the *Syrian Observatory for*

Human Rights, stated that ISIL had 50,000 fighters in Syria alone.[43] In September 2014, the CIA estimated ISIL's numbers were somewhere between 20,000 and 31,500 fighters between Iraq and Syria.[44] By November 2014, Fuad Hussein, the chief of staff to Kurdish President Massoud Barzani stated that ISIL had at least 200,000 fighters.[45] In January, 2015, Starr estimated that ISIL had between 9,000 and18,000 fighters.[46] The Russian chief of General Staff recently said that Russia estimates ISIL to have "70,000 gunmen of various nationalities."[47] In an online article that analyzes all the previous estimates, *Foundation for the Defense of Democracies* fellow Dr. Daveed Gartenstein-Ross estimates that based on the territory that ISIS controls between Iraq and Syria and the number of people within that territory, it is more likely that ISIS has close to 100,000 militants.[48]

Considering this discrepancy between numbers, it is nearly impossible to make any intelligent statement as to whether air strikes and drone strikes have decreased ISIS numbers at all. There have been several high-profile attacks where American air strikes and drone strikes have been conducted in conjunction. Drone strikes are used for precision targeting and targeted killing while air strikes are used to destroy large areas. However, the effects of drone strikes on ISIS numbers really cannot be determined until accurate numbers estimates are made. Therefore, there cannot even be a statement made as to whether American drone strikes and air strikes are actually working.

Concerning equipment, interestingly enough, ISIS has been using stolen American military equipment given to Iraqis. This American equipment was stolen from Iraq when ISIS entered Mosul and the Iraq army retreated from the city. The items left behind include U.S.-made armored Humvees, trucks, rockets, machine guns, and a helicopter. The list below includes the costs of each item and totals at around 656.4 million dollars.

- 2,300 Humvee armored vehicles @ $70,000 each. Total: $161 million
- 40 M1A1 Abram tanks @ $4.3 million each. Total: $172 million
- 52 M198 Howitzer mobile gun systems @ $527,337 each. Total: $27.4 million
- 74,000 army machine guns @ $4,000 each. Total: $296 million

ISIS has continued to use these items in their quest to create and conquer territory for the Islamic State. In fact, this military equipment has provided superior assistance to ISIS in their fight, particularly against local dissention groups.[49] The Department of Defense has kept track of the equipment that has been reclaimed from ISIS by coalition forces. As of March 29, 2016, this equipment is as follows in Table 6.2: *Operation Inherent Resolves Targets Damaged/Destroyed*.[50]

Table 6.2 Operation Inherent Resolve's Targets Damaged/Destroyed

Tanks	139
HMMWV's	374
Staging Areas	1,162
Buildings	5,894
Fighting Positions	7,118
Oil Infrastructure	1,272
Other Targets	6,820
Total	22,779

Source: Department of Defense, "Operation Inherent Resolve Targeted Operations Against ISIL Terrorists," March 29, 2016, accessed April 1, 2016, http://www.defense.gov/News/Special-Reports/0814_Inherent-Resolve.

Targeted Terrorists

Between Iraq and Syria, a few key ISIS figures have been assassinated mostly with some cooperation between drones and jets. In Iraq, Ayman al-Zarqawi, the founder of Al-Qa'ida in Iraq was found by drones and immediately assassinated by an F-16 with two 500lb precision bombs. One child, two of Zarqawi's wives and Sheikh Abd al-Rahman were killed by the same blast. Although Zarqawi was not killed initially, he died shortly after, succumbing to his injuries.[51] On August 18, 2015, Fadl Ahmed Abdullah al-Hiyali, the second in command of ISIS in Iraq, was killed in a drone strike near Mosul.[52] On April 3, 2016, a rocket expert, responsible for killing one Marine and wounding nine others, Jasim Khadijah was killed by drone strikes.[53]

In Al-Raqqah, Syria, on November 12, 2015, Jihadi John also known as Mohammed Emwazi was killed in a drone strike; John was responsible for beheading several journalists on camera and releasing it on the web.[54] On March 3, 2016, Abu al-Athir Amir al-Absi, governor of the ISIS Aleppo Province was killed either by the Americans or Russians in an airstrike, which could be a drone strike.[55] Omar al-Shishani, ISIS's Minister of War was killed in an airstrike on March 4, 2016.[56] ISIS commander Abd al-Rahman Mustafa al-Qaduli (also known as Haji Iman), was killed by a helicopter strike on March 25, 2016 along with Abu Sarah who was responsible for paying ISIS fighters.[57] Although American Special Operations had initially wanted to get al-Qaduli for questioning, a last minute decision was made to assassinate him instead.[58] On March 30, 2016, Abu Hija al-Tunisi, ISIS leader in Tunisia was assassinated by a drone strike in Syria.[59]

CONCLUSION

In conclusion, ISIS is like cancer. It has spread to Afghanistan, Saudi Arabia, Yemen, the Gulf States, Bangladesh, Lebanon, Jordan, the Gaza Strip, Egypt,

Libya, Tunisia, Algeria, Mali, and Nigeria in addition to other unknown countries. ISIS attacks have occurred in Europe, the United States, Canada, Australia, and Turkey. The group has continued to spread throughout the world. It does not appear that the United States has made tremendous headway in destroying ISIS at all.

Ashton Carter, the United States Secretary of Defense acknowledges that the United States has been pursuing the "cutting the head of the snake" objective for over two years. This strategy works to contain and destroy ISIS with air strikes and drone strikes while simultaneously keeping American boots off the ground. Carter states,

> We are systematically eliminating ISIL's cabinet. Striking leadership is necessary, but as you note, it's far from sufficient. Leaders can be replaced. However, these leaders have been around for a long time. They are senior, they're experienced, and so eliminating them is an important objective and it achieves an important result. But they will be replaced, and we'll continue to go after their leadership. [Airstrikes have destroyed] a significant quantity of improvised explosive devices and bomb-making equipment.[60]

Although it is true that airstrikes and drone strikes (sometimes just called airstrikes) have destroyed ISIS strongholds such as oil refineries and administration, it does not appear to be working. ISIS has committed major attacks in large European cities and they continue to be a threat several thousand miles away from their home bases. The migration of ISIS members to other countries to enlarge the caliphate cannot be considered success nor can the existence of sleeper cells. ISIS is always on the offensive. Although they have lost territory, they have reclaimed or gained more territory. When ISIS lost in Iraq, they simply rebuilt and moved into Syria. Air strikes and drone strikes have never put ISIS on the defensive. Putting ISIS on the defensive is one strategy that is required to beat ISIS.

In addition, from a tactical and military perspective ISIS is not stupid. They have surprised and gained respect from the American military and several other national militaries. ISIS uses car bombs as a frequent mode of support when they are on the offensive. Whether the driver volunteers for the suicide bombing or is chained to the steering wheel with a bomb that will detonate by remote control, ISIS has been known to use this method in the midst of a battle to strategically attack a foe.[61] In addition, ISIS uses copious IEDs when either leaving an area or when they are under attack.

There is one major reason why ISIS will continue to be successful in conquering more and more territory unless it is stopped. Despite its brutality, ISIS is fulfilling the civil services that the state such as Iraq and Syria has failed to provide. ISIS provides jobs, police services, legal services, food, water, electricity, and some medical care in addition to several other services that the state cannot provide. For this reason, ISIS will continue to expand as long as they are operating in failed or semi-failed states.

Although democracy is a lovely dream for countries like Iraq and Syria, it does not appear to be pursued by these country's citizens. Until democratic values and human rights (particularly children's and women's rights) are respected amongst the populous, it is unlikely that democracy is a pragmatic possibility. Unfortunately for the Middle East, strongman dictators appear to be the only realistic option in providing stability, progress, and peace. Perhaps one day, democracy will become an option.

NOTES

1 *The History Channel,* "9–11 Attacks," 2016, accessed March 18, 2016, http://www.history.com/topics/9-11-attacks.

2. Kyle Crichton, Gina Lamb, and Rogene Fisher Jacquette, "Timeline of Major Events in the Iraq War," *The New York Times*, 2003, accessed January 25, 2016, http://www.nytimes.com/interactive/2010/08/31/world/middleeast/20100831-Iraq-Timeline.html?_r=0#/.

3. Peter L. Hahn, *Missions Accomplished? The United States and Iraq since World War I?* (New York: Oxford University Press, 2012).

4. Daniel Trotta, "Iraq war costs U.S. more than $2 trillion: study," *Reuters*, March 14, 2013, accessed January 25, 2016, http://www.reuters.com/article/us-iraq-war-anniversary-idUSBRE92D0PG20130314.

5. For a detailed look at the costs of the Second War in Iraq including the long-term costs of veterans'' benefits, please see Watson Institute, *The Costs of War Project*, Brown University, 2016, accessed May 27, 2016, http://watson.brown.edu/costsofwar/.

6. Charles Lister, "Profiling the Islamic State," *Brookings Doha Center Analysis Paper*, November 2014, 12, accessed March 24, 2016, http://www.brookings.edu/~/media/Research/Files/Reports/2014/11/profiling%20islamic%20state%20lister/en_web_lister.pdf.

7. Ryan Evans, "From Iraq to Yemen: Al-Qa'ida's Shifting Strategies," *Combating Terrorism Center at Westpoint*," October 1, 2010, accessed January 12, 2016, https://www.ctc.usma.edu/posts/from-iraq-to-yemen-al-Qa'ida%E2%80%99s-shifting-strategies.

8. William McCants, *The ISIS Apocalypse, The History, Strategy, and Doomsday Vision of the Islamic State* (New York: St. Martin's Press, 2015).

9. Michael Weiss and Hassan Hassan, *ISIS Inside the Army of Terror* (New York: Regan Arts, 2015).

10. William McCants, *The ISIS Apocalypse, The History, Strategy, and Doomsday Vision of the Islamic State* (New York: St. Martin's Press, 2015).

11. Michael Weiss and Hassan Hassan, *ISIS Inside the Army of Terror* (New York: Regan Arts, 2015).

12. Charles Lister, "Profiling the Islamic State," *Brookings Doha Center Analysis Paper* 12, November 2014, accessed March 24, 2016, http://www.brookings.edu/~/media/Research/Files/Reports/2014/11/profiling%20islamic%20state%20lister/en_web_lister.pdf.

13. As cited in William McCants, *The ISIS Apocalypse, The History, Strategy, and Doomsday Vision of the Islamic State* (New York: St. Martin's Press, 2015), 85.

14. William McCants, *The ISIS Apocalypse, The History, Strategy, and Doomsday Vision of the Islamic State* (New York: St. Martin's Press, 2015).

Malcolm Nance, *Defeating ISIS, Who They Are, How They Fight, What they Believe* (New York: Skyhorse Publishing, 2016).

15. William McCants, *The ISIS Apocalypse, The History, Strategy, and Doomsday Vision of the Islamic State* (New York: St. Martin's Press, 2015).

16. Michael Weiss and Hassan Hassan, *ISIS Inside the Army of Terror* (New York: Regan Arts, 2015).

17. William McCants, *The ISIS Apocalypse, The History, Strategy, and Doomsday Vision of the Islamic State* (New York: St. Martin's Press, 2015).

Malcolm Nance, *Defeating ISIS, Who They Are, How They Fight, What They Believe* (New York: Skyhorse Publishing, 2016).

18. Luis Martinez, "Number of ISIS Fighters in Iraq and Syria Drops, Increases in Libya, US Official Says," *ABC News*, February 4, 2016, accessed March 24, 2016, http://abcnews.go.com/International/number-isis-fighters-iraq-syria-drops-increases-libya/story?id=36715635.

19. Luis Martinez, "Number of ISIS Fighters in Iraq and Syria Drops, Increases in Libya, US Official Says," *ABC News*, February 4, 2016, accessed March 24, 2016, http://abcnews.go.com/International/number-isis-fighters-iraq-syria-drops-increases-libya/story?id=36715635.

20. Charles Lister, "Profiling the Islamic State," *Brookings Doha Center Analysis Paper*, November 2014, 12, accessed March 24, 2016, http://www.brookings.edu/~/media/Research/Files/Reports/2014/11/profiling%20islamic%20state%20lister/en_web_lister.pdf.

21. Charles Lister, "Profiling the Islamic State," *Brookings Doha Center Analysis Paper*, November 2014, 12, accessed March 24, 2016, http://www.brookings.edu/~/media/Research/Files/Reports/2014/11/profiling%20islamic%20state%20lister/en_web_lister.pdf.

22. For information on drone strikes during the Iraq War, see Chris Woods, Chapter 4, *Sudden Justice* (Oxford: Oxford University Press, 2015).

23. Chris Woods, Chapter 4, *Sudden Justice* (Oxford: Oxford University Press, 2015), 76.

24. Chris Cole, "Drones in Iraq and Syria: What we know and what we don't," *Drone Wars UK*, July 11, 2014, accessed March 27, 2016, http://dronewars.net/2014/11/07/drones-in-iraq-and-syria-what-we-know-and-what-we-dont/.

25. Charles Lister, "Profiling the Islamic State," *Brookings Doha Center Analysis Paper*, November 2014, 12, accessed March 24, 2016, http://www.brookings.edu/~/media/Research/Files/Reports/2014/11/profiling%20islamic%20state%20lister/en_web_lister.pdf.

26. Chris Woods, "Strikes in Iraq," *Airwars.org*, 2016, accessed January 1, 2016, http://airwars.org/.

27. *Department of Defense,* "Operation Inherent Resolve Targeted Operations Against ISIL Terrorists," March 29, 2016, accessed April 1, 2016, http://www.defense.gov/News/Special-Reports/0814_Inherent-Resolve.

28. Chris Woods, "Strikes in Syria," *Airwars.org*, 2016, accessed January 1, 2016, http://airwars.org/.

29. Harleen Gambhir, "ISIS Regional Campaign, December 2015," *Institute for the Study of War*, January 11, 2016, accessed April 12, 2016, http://www.understandingwar.org/backgrounder/isiss-regional-campaign-december-2015.

30. The Guardian, "The ISIS papers: leaked documents show how Isis is building its state," *The Guardian*, December 7, 2015, accessed April 12, 2016, http://www.theguardian.com/world/2015/dec/07/leaked-isis-document-reveals-plan-building-state-syria.

31. *Department of Defense,* "Operation Inherent Resolve Targeted Operations Against ISIL Terrorists," March 29, 2016, accessed April 1, 2016, http://www.defense.gov/News/Special-Reports/0814_Inherent-Resolve.

32. Chelsea J. Carter, Tom Cohen, and Barbara Starr, "U.S. Jet Fighters, Drone Strikes, ISIS Fighters, Convoys in Iraq," *CNN*, August 9, 2014, accessed February 2, 2016, http://www.cnn.com/2014/08/08/world/iraq-options/.

33. *Airwars*, "Cause for Concern," August 2015, accessed April 17, 2016, http://airwars.org/wp-content/uploads/2015/08/airwars-cause-for-concern-civilians-killed-by-coalition.pdf.

34. *Iraq Body Count*, 2016, accessed March 27, 2016, https://www.iraqbodycount.org/.

35. Paul Wood and Richard Hall, "U.S. Killing More Civilians in Iraq, Syria Than it Acknowledges, *USA Today*, February 4, 2016, accessed April 12, 2016, http://www.usatoday.com/story/news/world/2016/02/02/us-killing-more-civilians-iraq-and-syria-than-acknowledges-globalpost/79686772/.

36. Syrian Network for Human Rights, "New Attacks by the International Coalition Forces Result in Losses and Civilian Victims," *Syrian Network for Human Rights*, February 13, 2016, accessed April 12, 2016, http://sn4hr.org/wp-content/pdf/english/68_civilians_killed_in_new_attacks_by_international_coalition_forces_en.pdf.

37. Syrian Network for Human Rights, "New Attacks by the International Coalition Forces Result in Losses and Civilian Victims," *Syrian Network for Human Rights*, February 13, 2016, accessed April 12, 2016, http://sn4hr.org/wp-content/pdf/english/68_civilians_killed_in_new_attacks_by_international_coalition_forces_en.pdf.

38. Syrian Network for Human Rights, "The Main Conflict Parties Who are Killing Civilians in Syria," *Syrian Network for Human Rights*, November 19, 2015, accessed April 12, 2016, http://sn4hr.org/wp-content/pdf/english/68_civilians_killed_in_new_attacks_by_international_coalition_forces_en.pdf.

39. Paul Wood and Richard Hall, "U.S. Killing More Civilians in Iraq, Syria Than it Acknowledges, *USA Today*, February 4, 2016, accessed April 12, 2016, http://www.usatoday.com/story/news/world/2016/02/02/us-killing-more-civilians-iraq-and-syria-than-acknowledges-globalpost/79686772/.

40. Paul Wood and Richard Hall, "U.S. Killing More Civilians in Iraq, Syria Than it Acknowledges, *USA Today*, February 4, 2016, accessed April 12, 2016, http://

www.usatoday.com/story/news/world/2016/02/02/us-killing-more-civilians-iraq-and-syria-than-acknowledges-globalpost/79686772/.

41. Jonathan Marcus, "Iraq and Syria opinion poll–the world's most dangerous survey?" *BBC*, September 9, 2015, accessed May 30, 2016, http://www.bbc.com/news/world-middle-east-34173549.

42. Riyadh Mohammed, "Islamic State Expands: Up to 100,000 People Have Joined, Experts Say," *Mashable*, August 26, 2014, accessed March 31, 2016, http://mashable.com/2014/08/26/100000-people-join-islamic-state/#aJF7hmnZ9sqN.

43. Al Jazeera and Agencies, "Islamic State 'has 50,000 fighters in Syria'," *al-Jazeera*, August 19, 2014, accessed March 31, 2016, http://www.aljazeera.com/news/middleeast/2014/08/islamic-state-50000-fighters-syria-2014819184258421392.html.

44. AFP Washington, "ISIS driving up fighter numbers in Iraq, Syria: CIA" *AFP*, September 12, 2014, accessed March 31, 2016, http://english.alarabiya.net/en/News/middle-east/2014/09/12/CIA-ISIS-has-20-000-to-31-500-fighters-in-Iraq-and-Syria.html.

45. Patrick Cockburn, "War with Isis: Islamic militants have army of 200,000, claims senior Kurdish leader," *Independent*, November 15, 2014, accessed March 31, 2016, http://www.independent.co.uk/news/world/middle-east/war-with-isis-islamic-militants-have-army-of-200000-claims-kurdish-leader-9863418.html.

46. Barbara Starr, "U.S. officials say 6,000 ISIS fighters killed in battles," *CNN*, January 22, 2015, accessed March 31, 2016, http://www.cnn.com/2015/01/22/politics/us-officials-say-6000-isis-fighters-killed-in-battles/.

47. Daveed Gartenstein-Ross, "How Many Fighters Does the Islamic State Really Have?" *War on the Rocks*, February 9, 2015, accessed March 31, 2015, http://warontherocks.com/2015/02/how-many-fighters-does-the-islamic-state-really-have/.

48. Daveed Gartenstein-Ross, "How Many Fighters Does the Islamic State Really Have?" *War on the Rocks*, February 9, 2015, accessed March 31, 2015, http://warontherocks.com/2015/02/how-many-fighters-does-the-islamic-state-really-have/.

49. Eric Pianan, "U.S. Shoots Itself In the Foot By Accidentally Arming ISIS," *The Fiscal Times*, June 4, 2015, accessed April 1, 2016, http://www.thefiscaltimes.com/2015/06/04/Fog-War-US-Has-Armed-ISIS.

50. *Department of Defense,* "Operation Inherent Resolve Targeted Operations Against ISIL Terrorists," March 29, 2016, accessed April 1, 2016, http://www.defense.gov/News/Special-Reports/0814_Inherent-Resolve.

51. Chris Woods, Chapter 4, *Sudden Justice* (Oxford: Oxford University Press, 2015), 82–83.

52. Malcolm Nance, *Defeating ISIS, Who They Are, How They Fight, What they Believe* (New York: Skyhorse Publishing, 2016), 36.

53. Stephen Kalin, "U.S. says it killed IS militant who killed Marine in Iraq," *Reuters*, April 4, 2016, accessed April 8, 2016, http://www.reuters.com/article/us-mideast-crisis-iraq-usa-idUSKCN0X00JS.

54. Dana Ford and Steve Almasy, "ISIS confirms death of 'Jihadi John,'" *CNN*, January 20, 2016, accessed April 16, 2016, http://www.cnn.com/2016/01/19/middleeast/jihadi-john-dead/.

55. Bill Roggio, "Islamic State's governor for Aleppo reportedly killed in airstrike," *Threat Matrix*, A Blog of the Long War Journal, March 4, 2016, accessed

April 16, 2016, http://www.longwarjournal.org/archives/2016/03/islamic-states-governor-for-aleppo-reportedly-killed-in-airstrike.php.

56. Eric Schmitt and Michael S. Schmidt, "Omar the Chechen, a Senior Leader in ISIS, Dies After U.S. Airstrike," *The New York Times*, March 15, 2016, accessed March 8, 2016, http://www.nytimes.com/2016/03/15/world/middleeast/omar-chechen-isis-killed-us-airstrike-syria.html.

57. Andrea Shalal and Susan Heavey, "Islamic State Second in Command, Likely Killed: U.S.," *Reuters*, March 26, 2016, accessed April 8, 2016, http://www.reuters.com/article/us-mideast-crisis-syria-isis-idUSKCN0WR12V.

58. Michael S. Schmidt and Mark Mazzeti, "A Top ISIS Leader Is Killed in an Airstrike, the Pentagon Says," *The New York Times,* March 25, 2016, accessed April 8, 2016, http://www.nytimes.com/2016/03/26/world/middleeast/abd-al-rahman-mustafa-al-qaduli-isis-reported-killed-in-syria.html?_r=0.

59. Arutz Sheva, "Drone strike kills another top ISIS commander," *Israel National News*, March 31, 2106, accessed Aril 8, 2016, http://www.israelnationalnews.com/News/News.aspx/210136#.Vwhodpvmphc.

60. Michael S. Schmidt and Mark Mazzeti, "A Top ISIS Leader Is Killed in an Airstrike, the Pentagon Says," *The New York Times,* March 25, 216, accessed April 8, 2016, http://www.nytimes.com/2016/03/26/world/middleeast/abd-al-rahman-mustafa-al-qaduli-isis-reported-killed-in-syria.html?_r=0.

61. David Kilcullen, *Blood Year, Islamic State and the War on Terror* (London: Hurst and Company, 2016), 144–145.

Conclusion

Is Targeted Killing Really Effective in the War on Terror?

In his Italian film, *5000 Feet is the Best*, Omer Fast portrays a scenario where an American family in Las Vegas goes on a short vacation and is attacked in the process by a drone strike. The film is meant to portray Pakistan in an American setting. Based on short interviews with a drone pilot from Creech Air Force Base, the audience hears quotes from the interview while watching the drone strike on the American family.[1] This literal play on words is the mock irony of American targeted killing in the Middle East.

Americans are secure in the fact that they will never be hit by a drone strike in their own country, that their houses will never be destroyed by a Hellfire missile, or that their children will never get hit by a wayward piece of shrapnel. As long as the United States is the sole superpower in the world, the American homeland is foreseeably safe, at least from drone strikes. However, the safety of the American citizens in some ways is sustained by the personal anguish of the Afghanis, the Iraqis, the Libyans, the Pakistanis, the Somalians, the Syrians, and the Yeminis and others throughout the world who are part of the drone war. It is these people that experience and suffer the collateral damage of the drone strikes.

In the introductory chapter, it was conveyed that the Conclusion would provide the following: a summation and comparison of all the data concerning targeted killing in the Middle East, an assessment of targeted killing as a counterterrorism policy in the Middle East (for both the United States and the Middle East), and the known fiscal costs of drone warfare. The concluding chapter will proceed as indicated.

TARGETED KILLING IN THE MIDDLE EAST, 2002–2015

In Table C.1: *Targeted Killing in the Middle East, 2002–2015*, the time frame, death rates, and deaths from drone strikes are presented. The death rates also include the terrorists who were targeted as it is unclear who is a terrorist and who is not. Unfortunately this table is incomplete in that Iraq and Syria do not have any data available on deaths and drone strikes in Afghanistan are not available from February 2013 to December 2014 as the United States' Air Force stopped publishing their drone strike numbers. The deaths from drone strikes in Afghanistan are also not available during these years. Afghanistan was the only country whose drone strikes numbers were published by the United States' Government. Think tanks like *The Long War Journal* and *The Bureau of Investigative Journalism* are the only reason the data for the rest of the countries is even available. The data used in this book was taken from the most complete websites available and is not combined from various sources. For Pakistan and Afghanistan, there are a few sources concerning drone strikes numbers and their estimates vary according to source, but again various sources are not presented because that data is not the most complete concerning the time frame. In addition, many drone strikes are never cataloged or detected in rural areas of the Middle East. If all of the data were included, these numbers would be much higher. If half of 2016 were included, the deaths per drone strike would be much higher as targeted killing has increased astronomically in 2016 as the individual chapters have elaborated upon. A conservative estimate of the number of deaths from drone strikes from 2002 until now in August 2016 in all seven Middle Eastern countries would include over 10,000 people. A look at the terrorists killed in each country would add up to a few hundred, depending on what the definition of "terrorist" is. An al-Qa'ida leader is definitely a terrorist but a foot soldier or truck driver for al-Qa'ida is a little more difficult to ascertain.

An Assessment of Targeted Killing in the Middle East, 2002–2015

In the Introduction, it was stated that an assessment of targeted killing in the Middle East would be based on the following:

1. Effective counterterrorism requires a decrease in terrorism in the countries where drone strikes occur whether categorized as terrorist attacks or suicide terrorist attacks. An increase in drone strikes should then have a decrease in terrorist attacks in each country.
2. Effective counterterrorism does not create blowback in which radicalization occurs, anti-Americanism is created, and terrorist group recruitment and empathy increases.

Table C.1 Targeted Killing in the Middle East, 2002–2015

Country	*Time Frame of Drone Strikes*	*Deaths Per Drone Strike*	*Deaths from Drone Strikes*
Afghanistan	Began January 2009	10.83 (2015 only)	1,517 (2015 only)
Iraq	Began August 2014	Data unavailable	Data unavailable
Libya	March to October 2011	7.3	1,269
Pakistan	Began June 2004	6	2,500
Somalia	Began June 2011	6 (2011–2015)	126
Syria	Began September 2014	Data unavailable	Data unavailable
Yemen	Began November 2002	7.3 (2002–2015)	1269

3. Effective counterterrorism has a collateral damage rate of less than 10 percent. This means from a human being perspective that for every terrorist that is killed, there must be 0.1 civilians killed or for every ten terrorists that are killed, approximately one civilian is killed.

It can be stated that drone strikes and targeted killing in the Middle East are not working for the following reasons:

1. Terrorist attacks and suicide bombings have increased in Afghanistan, Iraq, Pakistan, Somalia, Syria, and Yemen since the drone strikes began. Although Somalia is currently experiencing very few drone strikes, this may change if al-Shabaab becomes more powerful or if ISIS unites with al-Shabaab in Somalia. The War on Terror was supposed to decrease terrorism around the world and the numbers of terrorists. With more data, the quantitative methodological case can probably be made that drone strikes are increasing terrorist attacks or at least contributing to their increase. However, the data is simply not available to do this analysis correctly across the board as of yet. Either way, whether the drone strikes are responsible for more terrorist attacks or not, drone strikes are definitely not decreasing terrorist attacks.
2. The number of people killed by drone strikes will continue to increase as has been seen in the second term of the Obama Administration. Although the Obama Administration has purported to be decreasing civilian casualty rates, this does not appear to be true in 2016 across the board. It is likely that the next person elected president in 2016 will most likely, regardless of party, continue to conduct drone strikes. In article written by Mattea Kramer,[2] the author found that every presidential candidate including Donald Trump and Hillary Clinton are pro-drone; the next administration will pursue a continuation of Obama's targeted killing policy. The problem is that too many civilians are being killed by drone strikes. If the Air

Force's collateral damage rate is 0.1 person per drone strike or one person is killed per ten drone strikes, the United States is not even remotely close to accomplishing their collateral damage goals.

3. As a counterterrorism method, drone strikes have not decreased the strength of ISIS in Iraq and Syria nor have the other counterterrorism methods the United States has used defeated ISIS. ISIS is still expanding their territory and every day includes another news story on the fight with ISIS. If ISIS was being defeated by drone strikes, they would have never advanced into Syria and most importantly, they would have never made it to Baghdad in the beginning. If ISIS were being defeated, they would never attack Western homelands in Belgium, Paris, or the United States in addition to several other countries. Although the United States' military publishes constant messages stating that America is beating ISIS, this does not appear to be the case at all but propaganda. ISIS has had more than ninety attacks in twenty-one countries.[3]
4. The number of people being radicalized by drone strikes is unknown but the chapters provide evidence, even though it is anecdotal that something is going on here. People obviously in large majorities do not support or like the drone strikes. The available public opinion polls support this. The drone strikes are creating anti-Americanism. This American heavy-handed approach to counterterrorism does not work but makes people angry. Then there are many people who talk about joining terrorist groups because their family member or in some cases, their entire family was killed by a drone strike because the United States made a mistake. We have already seen terrorists attack the United States in retaliation for drone strikes in Pakistan and Afghanistan. This trend will continue. The terrorist groups do not seem to be decreasing in numbers so it is obvious that they are receiving recruits for various reasons. A body count approach does not work as there are more than enough people that join the terrorist group to sustain it.
5. Although it is true that some high-level terrorists have been killed by drone strikes, many of the terrorists that are killed are low-level people. Drone strikes should be used rarely to kill high level terrorists like Osama bin Laden who are possible founders of the organization. The problem with killing leadership in a terrorist organization is that there appears to be someone always waiting in the wings. The supply of recruits is endless and although the new leader may require some training, somehow the next person always seems to get the job done, sometimes with a vengeance. In an article by Rosa Brooks,[4] she states that decapitating leadership does not work and does not pass a cost-benefit analysis. Today's terrorist groups are decentralized making leadership less important and almost irrelevant to the terrorist organization when a leader is killed. This technique may work in traditional national terrorist organizations like the FARC but international

terrorists organizations that are a relatively new phenomenon like ISIS do not seem to be affected much by leadership decapitation.

6. Drone strikes are not producing better outcomes for the countries in any of these case studies. Both Afghanistan and Iraq are currently political disasters with ISIS still in Iraq. Somalia, Syria, and Yemen are all in the midst of civil wars with Somalia and Yemen being failed states. Syria is fighting ISIS too. Pakistan is probably doing the best but is still a violent country. In Libya, the intervention led by the United States but conducted by NATO did not lead to a better outcome. Obviously, airstrikes. were used in the NATO intervention but Libya is worse off now than under Qaddafi. These drone strikes have not produced any major positive outcomes in these countries, save the deaths of some terrorists.

Concerning the United States, it is not clear if the United States is safer from terrorist attacks. The reasons for this are as follows: The foiled terrorist attacks are not published by the United States' Government and we are rarely given information if an attack is stopped. So if drone strikes were creating less terrorist attacks, we would not know. In addition, the United States has had some terrorist attacks like the Boston bombings occur, some inspired or linked to ISIS. This does point to the fact that drone strikes are not necessarily making the United States safer from ISIS or other terrorist groups.

Recently in 2016 the Obama Administration has allowed less than 2500 Syrian refugees into the United States.[5] Many Americans are afraid of these Syrian refugees as it is likely that there are some ISIS members or sympathizers within this group of people. In addition, around 50 percent were men with a large percentage of them being military-aged.[6] Obviously it is possible that some terrorist attacks may occur as a result of letting these refugees into the United States.

But What about the Military?

Although this book is about the problems with drone strikes, there are several other great uses for drones. One of the best uses of drones is to protect and assist the military. Drones have made it much easier to attack and find the enemy. Michael Waltz, a Special Forces commander in Afghanistan states that

> Drones brought a tremendous capability for us to see things and make decisions that were impossible only a few years ago. In 2005, if I called an airstrike as a response to someone firing at me from a compound, they would probably have destroyed the entire compound. Now, if I have access to a drone I can see that, even though we're being attacked from that compound, there are also seven

> women inside. That information allows for a different decision and a different way of managing the threat.[7]

Waltz also talks about the time a drone saved his life when he was under attack from a three-sided ambush. The drones were able to see more of the enemy coming and so Waltz was forced to retreat. There are also times when drones are able to check out the territory ahead allowing safe passage for troop movements.[8] Drones are highly useful in many capacities including protecting troops and surveillance operations.

The author is making the point that drones should not become the only counterterrorism tactic in the War on Terrorism. The Obama Administration has used drones to assassinate terrorists at will with the refusal of the media to pay attention to what the president is doing. Intelligence should be the main tactic in the War on Terror, not killing people with drones. Targeted killing should only be used when there is an imminent threat or an emergency not on a daily basis. Dan Hampton, a former drone pilot makes the following points.

> Then there is the probe of mentality. With a decade of permissive skies over Afghanistan and Iraq, some flawed conclusions are being reached. The love affair with stand-off weapons and unmanned aerial vehicles are perfect examples. Once again, those who are too old or who were never good enough for combat continue to advocate the replacement of manned aircraft. These are the same guys who said we'd never strafe and we'd never drop bombs on SAM sites. But combat pilots know that as long as ground troops are fighting, they'll need close air-support and Weasels. . . I know that UAVs, in their present form, would never survive in any threatening environment with MiGs or SAMs or Triple-A. When generals began insisting that fighters fly escort for the things, then we were at the edge of the abyss. We went over into the abyss when the Air Force made a noncombat officer the Chief of Staff.[9]

THE FISCAL COSTS OF DRONE WARFARE

Although the United States' military has embraced drones as the highly anticipated, low casualty, warfare weapon of the future, a thorough cost analysis of drone warfare has not been performed. The War in Afghanistan cost over one trillion dollars, not including unknown future costs associated with veteran benefits over the long term.[10,11] The United States spent over 1.7 trillion with an additional 490 billion in benefits that will be paid to war veterans over the next several decades. The total cost of the Iraq War came to around 2.19 trillion with included estimated veteran benefits.[12,13]

This part of the Conclusion focuses on the costs of drone warfare. This attempt at quantifying the drone war should not be taken as the exact price of the drone war but is meant to provide a possible estimate of the drone wars. The purpose is to try to understand what is coming out of the American people's pockets. Niether every cost nor every mistake can be accounted for. However, this analysis does include every quantifiable known cost of the drone wars in Afghanistan, Iraq, Libya, Pakistan, Somalia, Syria, and Yemen. The Philippines and Nigeria are also a target for drone warfare although the numbers of these strikes are not available and the details are relatively unknown at this point in time.

This section concerning the costs of drones has six separate sections. The first section considers the price of the American drone arsenal and the cost of the actual drones the United States has acquired. The second section will look at the cost of all drone strikes up until the end of 2015. Section three will identify the costs of flying time and maintenance for three of the most widely used drones by the American military. Section four will examine the costs of programs for training pilots based on data from Holloman Air Force Base, the major training facility for drone pilots, and a former pilot's estimation of training costs. Section five will explore the fiscal budgets for research and development concerning drones. Lastly, section six consists of the estimates of the costs of all the drones than have been crashed and destroyed by the United States.

The Cost of the American Drone Arsenal

In Table C.2: *American Military Drones Projected 2017*, the approximate costs of the arsenal of American drones in 2017 is presented. The sheer number of drones is larger than the 2016 arsenal but not by much. Matthew Aid an author and intelligence expert published a report from the Pentagon concerning the expected number of drones the United States would own in 2017. This list is neither exhaustive nor exact. Instead, it is an estimate of the costs of the drones the United States will own in 2017. Some of the drones are owned by more than one branch of the military so the total number of drones that the United States owns, regardless of the branch of service are given. Not including drones ownership after 9–11 and before 2017, and drones that have been crashed etc., the total value of the American arsenal of drones in 2017 will be approximately 21.5 billion.

The Costs of American Drone Strikes

Presented below in Table C.3: *The Costs of Hellfire missiles in American Drone Strikes, 2002–2015* are all the costs of Hellfire missiles concerning known American drone strikes in the War on Terror. The first strike was in Yemen in 2002. Since 2002, the United States has had numerous strikes in Afghanistan,

Table C.2 American Military Drones Projected 2017

Name	*Maker*	*Cost*	*Estimated Number*	*Total*
Global Hawk	Northrop Grumman	225,000,000	15	3,375,000,000
Fire Scout	Northrop Grumman	18,000,000	37	666,000,000
Reaper	General Atomics	14,000,000	256	3,584,000,000
Predator	General Atomics	5,000,000	110	550,000,000
Scan Eagle	Boeing and Insitu	3,000,000	122	366,000,000
Raven	AeroVironment	35,000	7074	247,590,000
Hunter	Northrop Grumman	3,000,000	45	135,000,000
Grey Eagle	General Atomics	21,500,000	152	3,268,000,000
Shadow	AAI Corporation	15,500,000	460	7,130,000,000
Triton	Northrop Grumman	120,689,000	13	1,568,957,000
Blackjack	Boeing	559,000,000	104	559,000,000
Avenger	General Atomics	15,000,000	4	60,000,000
Total				21,509,547,000

Source: Simon Rogers, "Drones by country: who has all the UAVs?" The Guardian, August 3, 2012, accessed February 4, 2016, http://www.theguardian.com/news/datablog/2012/aug/03/drone-stocks-by-country.
Key Data, "Predator RQ-1 / MQ-1 / MQ-9 Reaper UAV, United States of America," air-force-technology.com, 2016, accessed February 4, 2016, http://www.airforce-technology.com/projects/predator-uav/.
Spencer Ackerman and Noah Shachtman, "Almost 1 in 3 U.S. War Planes in a Robot," Wired, January 9, 2012, accessed February 4, 2016, http://www.wired.com/2012/01/drone-report/.
Matthew Aid, "How Many Unmanned Drones is the US Flying?" Matthewaid.com, August 20, 2012, accessed February 5, 2016, http://www.matthewaid.com/post/29822072373/how-many-unmanned-drones-is-the-us-flying.

Iraq, Libya, Pakistan, Somalia, Syria, and Yemen in an effort to kill terrorists and terrorist training sites. This data has been compiled from numerous think tanks and sites including *The Bureau of Investigative Journalism, The Long War Journal*, and the American Air Force. The cost of one Hellfire missile is approximately 70,000 dollars.[14] Each strike uses at least one, usually two Hellfire missiles.[15] These are the typical weapons used in a drone strike. The estimate below is a conservative estimate of the costs of Hellfire missile use. Each Predator, the typical drones used in these attacks, can hold two Hellfire missiles. The Reaper, which has been used since 2007, can hold up to four Hellfire missiles. The costs of Hellfire missile use alone since 2002 has been approximately 4 million and this is a conservative estimate.

The Costs of Flying Time and Maintenance

In Table C.4: *The Costs of Flight*, flight costs are presented for the total hours of flight per hour (CPFH) for the most widely used drones. There is no universal calculation formula for the costs of flight per hour. The CPFH that are given below are calculated by the Department of Defense. The CPFH typically includes maintenance costs, asset utilization costs, and military personnel costs. Direct operating costs may also include such items as maintenance,

Table C.3 The Costs of Hellfire Missiles in American Drone Strikes 2002–2015

Country	*Time Frame*	*Number Drone Strikes*	*Cost of One Hellfire Missile*	*Total*
Afghanistan[1]	Jan 2009–Dec 2015	19102	70000*2	2,674,280,000
Iraq[2]	Jan 2008–Dec 2015	5906	70000*2	826,840,000
Libya[3]	2011	105	70000*2	14,700,000
Pakistan[4]	June 2004–Dec 2015	429	70000*2	60,060,000
Somalia[5]	June 2011–Dec 2015	19	70000*2	2,660,000
Syria[6]	Sept 2014–Dec 2015	3188	70000*2	446,320,000
Yemen[7]	Nov 2002–Dec 2015	274	70000*2	38,360,000
Total		29023		4,063,220,000

[1]For 2009–2012 (Data below was pulled by the US Air Force in 2015): Noah Shachtman, "Military Stats Reveal Epicenter of US Drone War," *Wired.com*, November 9, 2012, accessed February 23, 2016, *http://* www.wired.com/2012/11/drones-afghan-air-war/; American Drone strikes for 2013 and 2014 in Afghanistan are not available because they are still classified by the United States Air Force; For 2015: Jack Serle and Payenda Sargand, "Get the data: A list of US air and drone strikes, Afghanistan 2015," *The Bureau of Investigative Journalism*, February 12, 2015, accessed August 13, 2015, https://www.thebureauinvestigates.com/2015/02/12/us-drone-war-afghanistan-list-american-air-strikes-2015/.

[2]For early Iraq data, see Chris Woods and Alice K Ross, "Drone Warfare Revealed: US and Britain launched 1,200 drone strikes in recent wars," *The Bureau of Investigative Journalism*, December 4, 2012, accessed February 23, 2016, https://www.thebureauinvestigates.com/2012/12/04/revealed-us-and-britain-launched-1200-drone-strikes-in-recent-wars/; Chris Woods, "Iraq," Airwars.org, 2016, accessed February 23, 2016, http://airwars.org/.

[3]Chris Woods and Alice K Ross, "Drone Warfare Revealed: US and Britain launched 1,200 drone strikes in recent wars," *The Bureau of Investigative Journalism*, December 4, 2012, accessed February 23, 2016, https://www.thebureauinvestigates.com/2012/12/04/revealed-us-and-britain-launched-1200-drone-strikes-in-recent-wars/.

[4]*The Bureau of Investigative Journalism*, "Get the Data: Drone Wars Pakistan," *The Bureau of Investigative Journalism*, 2016, accessed February 23, 2016, https://www.thebureauinvestigates.com/category/projects/drones/drones-graphs/.

[5]*The Bureau of Investigative Journalism*, "Get the Data: Drone Wars Somalia," *The Bureau of Investigative Journalism,* 2016, accessed February 23, 2016, https://www.thebureauinvestigates.com/category/projects/drones/drones-graphs/.

[6]Chris Woods, "Syria," Airwars.org, 2016, accessed February 23, 2016, http://airwars.org/

[7]*The Bureau of Investigative Journalism*, "Get the Data: Drone Wars Yemen," *The Bureau of Investigative Journalism,* 2016, accessed February 23, 2016, https://www.thebureauinvestigates.com/category/projects/drones/drones-graphs/.

Table C.4 The Costs of Flight until 2013

Drone	*Cost Per Hour*	*Total Flying Time*	*Total*
Global Hawk	49,089	225,000	11,045,025,000
Reaper	4,762	1,000,000	4,762,000,000
Predator	3,679	1,000,000	3,679,000,000
Total			19,486,025,000

Source: Friends Committee on National Legislation, "Understanding Drones," Friends Committee on National Legislation, 2016, accessed February 5, 2016, http://fcnl.org/issues/foreign_policy/understanding_drones/.

Marks Thompson, "Costly Flight Hours," Time, April 2, 2013, accessed February 5, 2016, http://nation.time.com/2013/04/02/costly-flight-hours/.

Northrop Grumman, "Firescout Unprecedented Persistent Situational Awareness," 2016, accessed February 5, 2016, http://www.northropgrumman.com/Capabilities/firescout/Pages/default.aspx.

Northrop Grumman, "Global hawk," 2016, accessed February 5, 2016, http://www.northropgrumman.com/capabilities/globalhawk/Pages/default.aspx.

fuel, and lubricants. Costs are typically dependent upon the number of hours flown. "To this may be added hanger, maintenance and personnel costs (direct and indirect) including operators, mechanics and administrators of the program. Some cost estimates may include depreciation and the present value of the costs per contract divided by the average contract length in order to estimate total operational costs."[16]

The "Total Flying Time" is the record of the number of hours that the drone has reached across the board regardless of how many drones were used to achieve the total flying time. Both the Reaper and the Predator reached one million hours within the same time frame in October 2013. Obviously, these estimates are not relatively recent but they give a conservative estimate as to how much money has actually been spent in deploying the drones.[17] The total CPFH so far for the three most widely used drones by the American military is around twenty million dollars since the drones were first deployed.

The Costs of Training Pilots

The primary location for drone training is Holloman Air Force Base is Alamogordo, New Mexico. At Holloman, UAV are allowed to fly approximately twelve hours a day through the vast air space that Holloman is given access to. Holloman has over 60,000 acres of land and an altitude of over 4000 feet. Even countries that buy UAV from the United States will have or have had training at Holloman Air Force Base.[18]

Training is primarily led by CAE, originally a Canadian company, which has been providing training on and off since 1998. There are approximately 100 CAE instructors at Holloman who make up 35–40 percent of the total instructors. The rest are USAF personnel. Under the guidance of the 49th Operations Group, the squadrons responsible for training pilots on UAV are the 6th Reconnaissance, the 9th Attack, the 16th Training, and the 29th Attack. Creech Air Force Base outside Las Vegas Nevada, previously responsible for Predator training is now only responsible for the launch and recovery aspects of UAV. "Because they are so difficult to fly, when the students come through here, they are mission qualified to do the mission in the airplane, but they are not qualified to launch it, and they are not qualified to land it," stated one UAV trainer at Holloman Air Force Base.[19]

Holloman graduated approximately 800 pilots and sensor operators in 2015. This is an increase of eighty pilots from 2014. Approximately two-thirds of the training takes place on the Reapers while the rest of the training occurs on the Predators. The reason for the disparity is that the MQ-1 Predator will be phased out by 2018. There are also two smaller UAV training sites that are owned by the National Guard, one located at March Air Force Base in California and the other at Hancock Field in Syracuse, New York.

These bases train approximately 80–100 students per year although they are primarily for National Guard soldiers.[20]

The training for a pilot and his sensor operator (co-pilot) is approximately four and one-half months, which is included in the whole process typically taking around one year. Prior to Holloman, trainees receive instruction and undergraduate training at Pueblo, Colorado at the Randolph Air Force Base. Trainees are also screened at the Randolph Air Force Base before their transfer to Holloman. By the time students graduate from Holloman, approximately 90 percent of students will have dropped a GBU-12 Paveway bomb from a Predator and occasionally a Lockheed Martin Hellfire missile, which is the primary killing tool for targeted killing. Not all trainees come to Holloman straight from school; some UAV pilots are actually USAF jet pilots who have decided to train on UAV.[21,22] In Figure C.1: *UAV Training Process*,[23] this cycle is visualized concerning where each particular type of training takes place.[24]

In Table C.5: *The Costs of Drone Pilot Training*, the number of pilots trained per year at Holloman Air Force Base are provided.[25] The costs of training are taken from the memoirs of a drone pilot, Lt. Col Matt Martin.

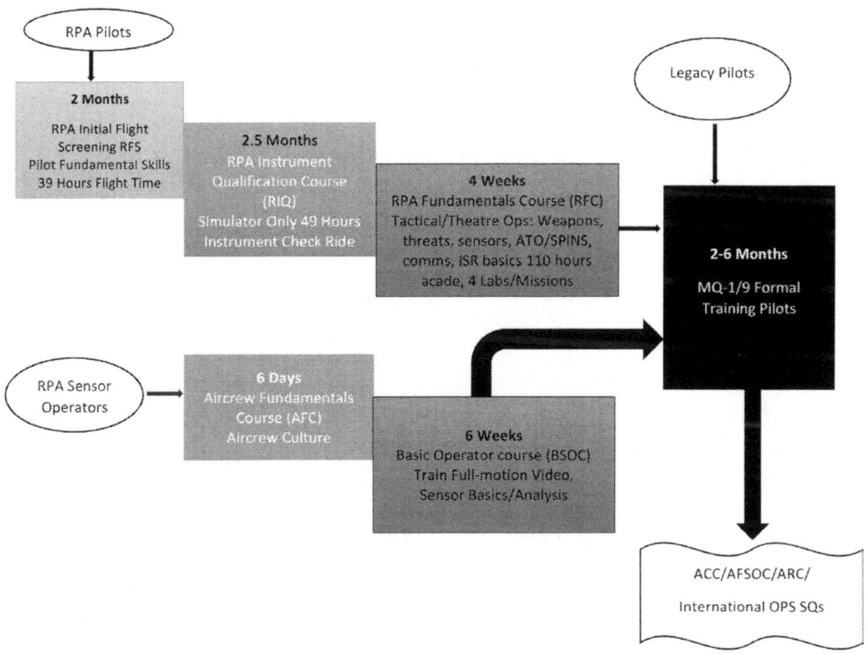

Figure C.1 Undergraduate RPA Training (Pipeline). *Source*: Arlan Ponder, Director Media Relations, "RPA Mission Brief," September 10, 2015, Holloman Air Force Base, Alamogordo, New Mexico.

Table C.5 The Costs of Drone Pilot Training

Year	*Number of Pilots Trained*	*Cost of Training*	*Total Cost*
2009	136	250,000	34,000,000
2010	293	250,000	73,250,000
2011	451	250,000	112,750,000
2012	566	250,000	141,500,000
2013	678	250,000	169,500,000
2014	716	250,000	179,000,000
2015	603	250,000	150,750,000
2016	818	250,000	204,500,000
Total			1,065,250,000

Lt. Col. Matt Martin stated that during his twelve weeks of training at Creech Air Force Base,

> Students scraped up three aircraft tailboards at a replacement cost of ten grand each; crashed one aircraft, destroying it, to the tune of about four million; burned up thousands of gallons of aviation fuel; and shot off eight Hellfire missiles at sixty-eight thousand dollars per. I figured Uncle Sam spent a cool quarter million dollars training me to be a Predator pilot.[26]

As previously mentioned although Creech used to do more training of drone pilots, Holloman Air Force Base is now the primary training facility. Considering, the costs of the aircraft that is damaged per student, these estimates are again most likely conservative. As can be seen in Table C.5: *The Costs of Drone Pilot Training*, there are approximately one billion since 2009 when the first statistics are available.

Research and Development Budget

Although many components of the budget that will be mentioned in this section have already been accounted for in one of the previous sections, it is worth discussing the budget measures for the fiscal year 2016 concerning the Department of Defense's drone spending. In 2016, the Pentagon was allocated 573 billion for the base account and 50.9 billion for the Overseas Contingency Operations (OCO), which is part of the drone warfare and operations. Most of the projects below were included in the budget and some new ones were added by the Senate for the final budget.[27]

The Air Force will receive twenty-six new MQ-9 Reapers totaling 821 million for the Air Force. 82.5 million will be taken for OCO and sixty-nine million for extending the ranges of UAV. The Air Force plans to spend

769.1 million on 5859 Hellfire missiles for the Predators and Reapers. The Air Force will also increase its incentive pay for drone pilots as the USAF is currently experiencing a number of drone pilots who are quitting their careers for various reasons. One purpose for this loss of drone pilots is PTSD.[28]

Also part of this budget is the Navy's expectation to purchase three Tritons for 548.8 million. Seven Blackjack drones will be purchased by the Navy and Marines. This will include three Blackjacks for the Navy at fifty-five million and four Blackjacks for the Marines at 84.9 million. 3.8 million will be devoted to upgrades on the Shadow and 3.4 million for the Raven upgrades. The Navy will also purchase two Fire Scouts at 120 million or sixty million per Fire Scout. The army plans to buy seventeen Gray Eagles for 383 million total and will spend 89.7 million on upgrades to the Shadow. About 1.325 million will be expended on Apache helicopters.[29]

Part of the budget will include research and development spending. The USAF will spend 51.3 million to improve ground control stations for the drones and another forty-seven million for upgrades. Another 208.1 million will be spent on development of the Global Hawk. Around 27.578 million will be used by the USAF on the control interface for UAV. 6.602 million will be devoted to the ability of the Air Force pilots to control multiple UAV at one time. Around 24.245 million dollars will be allocated for developing the airborne control teams.[30]

The Navy and Marines will also spend millions of dollars on research and development. The Navy will use 134.7 million to review the Unmanned Carrier-Launched Airborne Surveillance and Strike (UCLASS) development program. 26.885 million will be spent on lasers for the Navy. Thirty-four million will be allocated for researching an Unmanned Surface Vehicle. The Navy will spend twenty-five million on an Unmanned Undersea Vehicle. Around 17.392 million will be used to develop the Knifefish Surface Mine Countermeasure Unmanned Undersea Vehicle.[31]

Lastly, the army will spend 40.372 million on developing a Tactical Unmanned Ground Vehicle, The Common Robotic System. 8.426 million will be spent on Robotics Technology research. The army will spend 7.592 million on Intelligent Systems Technology Research. Around 4.242 million will be used on human-robot interaction including an additional 7.233 million for research on robotic anatomy. Around 6.489 million will be used for the development of collaboration of intelligence technology in unmanned vehicles. 2.029 million will be used on the creation of algorithms which will help soldiers predict signature content. About 5.841 million will be spent on research micro autonomous systems technology and 1.762 million to decrease energy used by micro/small scale drones to copy behaviors found in the biological environment. Micro autonomous Systems Technology will require an additional 7.299 in spending.[32]

For construction, the USAF will use fifty million for an airfield in Agadez, Niger and RAF Croughton in England will receive 130.615 million for the creation of a new satellite communication facility. The Navy will use 40.641 million to upgrade a Naval Air Station in Italy for the Triton and 8.296 Million will be used to construct a Mission Control Systems facility in Jacksonville, Florida. A naval base in Ventura County, California will require 2.974 million for a Triton training facility. The Marines will also need 29.657 million to create an Unmanned Air System Facility at Cherry Point, North Carolina.[33]

Since there is a possibility to double count some of the budget money used in this section, particularly on the purchases of the UAV, this money will not be used in the tally of the final costs in the comparison section. However, the budget was worth mentioning in this section as it gives the reader a sense of the amount of money that is being used in the development, procurement, and research on UAV technology.

Great! Now You Broke It!

The following Table C.6: *American Drone Crashes* is a list of drones that have been crashed by the United States. This list was compiled by Drone Wars UK,[34] an independent non-profit that was created to track the usage and crashing of drones. The name of the drone and the country that crashed the drone are listed since 2007 when the non-profit started tracking crashes. This is not a complete list of drone crashes. The cost of American drone crashes since 2007 has been somewhere around three billion dollars.

A COST COMPARISON

In Table C.7: *The Total Costs of Drone Warfare*, the total of all drone warfare costs provided in this section are given. The total cost of American drone warfare comes to around fifty billion. Of course, there are most likely many more costs which are not included in this figure. However it is safe to say, regardless of the costs that are not included, that the War in Iraq and the War in Afghanistan were much more expensive than the American drone war that has been occurring since 2002. The drone war is approximately twenty times cheaper than the War in Afghanistan and approximately thirty-four times cheaper than the War in Iraq. As recited earlier, the War in Afghanistan was approximately one trillion and the War in Iraq was approximately 1.7 trillion.

From a time perspective, the War in Afghanistan will last approximately fourteen years if the Obama Administration does remove all troops by the end of 2017 as planned. The War in Iraq will have lasted approximately six years. The drone war has been occurring since 2002 and will likely continue much

Table C.6 American Drone Crashes

Year	*Crash Victims*	*Costs*	*Total*
2007	A160T-1	4,000,000	94,500,000
	MQ Predator-7	35,000,000	
	MQ-9A Reaper-1	14,000,000	
	MQ5B-Hunter-5	20,000,000	
	RQ-1C Warrior-1	21,500,000	
2008	MQ Predator-12	60,000,000	758,600,000
	MQ-9A Reaper-3	42,000,000	
	MQ5A and B-Hunter-3	9,000,000	
	RQ-1C Warrior-3	645,000,000	
	RQ-1L Gnat-1	2,600,000	
2009	MQ Predator-13	65,000,000	382,100,000
	MQ-9 Reaper-4	56,000,000	
	MQ5A and B-Hunter-4	12,000,000	
	RQ-1C Warrior-1	21,500,000	
	RQ-1L Gnat-1	2,600,000	
	RQ-4A-Global Hawk-1	225,000,000	
2010	MQ Predator-10	50,000,000	111,500,000
	MQ-9 Reaper-2	28,000,000	
	MQ5A and B-Hunter-4	12,000,000	
	RQ-1C Warrior-1	21,500,000	
2011	MQ Predator-14	70,000,000	403,000,000
	MQ-1C Gray Eagle-1	21,500,000	
	MQ-9 Reaper-4	56,000,000	
	MQ5A and B-Hunter-1	3,000,000	
	MQ-170 Sentinel-1	6,000,000	
	RQ-1C Warrior-1	21,500,000	
	RQ-4A-Global Hawk-1	225,000,000	
2012	MQ Predator-10	50,000,000	541,500,000
	MQ-1C Gray Eagle-6	129,000,000	
	MQ-8 Fire Scout-3	54,000,000	
	MQ-9 Reaper-4	56,000,000	
	MQ5A and B-Hunter-2	6,000,000	
	RQ-1C Warrior-1	21,500,000	
	RQ-4A-Global Hawk-1	225,000,000	
2013	KMAX-1	4,000,000	221,400,000
	MQ Predator-8	40,000,000	
	MQ-1C Gray Eagle-3	64,500,000	
	MQ-8 Fire Scout-1	18,000,000	
	MQ-9 Reaper-5	70,000,000	
	MQ5A and B-Hunter-1	3,000,000	
	RQ-1C Warrior-1	21,500,000	
	S100 Camcopter-1	400,000	
2014	MQ Predator-10	50,000,000	92,000,000
	MQ-9 Reaper-3	42,000,000	
2015	MQ Predator-8	40,000,000	69,500,000
	MQ-9 Reaper-1	14,000,000	
	RQ-7B Shadow-1	15,500,000	
Total			**2,674,100,000**

Source: Drone Crash Database, Drone Wars UK, 2015, accessed February 6, 2016, http://dronewars.net/drone-crash-database/.

Table C.7 The Total Costs of Drone Warfare

Section	Total Cost
Table C.2: American Military Drones Projected 2017	**21,509,547,000**
Table C.3: The Costs of Hellfire Missiles in American Drone Strikes 2002–2015	**4,063,220,000**
Table C.4: The Costs of Flight until 2013	**19,486,025,000**
Table C.6: The Costs of Drone Pilot Training	**1,065,250,000**
Table C.7: American Drone Crashes	**2,674,100,000**
Total:	**48,498,142,000**

longer than the Iraq War or Afghanistan War. The current drone war is costing somewhere around four billion a year (if we divide fifty billion by thirteen years), which is relatively cheap compared to traditional warfare or at least recent traditional warfare. At this rate of four billion per year that has been calculated, it will take approximately 238 years to reach the spending of the Afghanistan War and approximately 413 years to reach the same spending as the Iraq War. With that being said, the author can safely say, regardless of the data that has not been collected, that the costs of drone warfare are tremendously cheaper than the costs of traditional warfare.

The major component that was not included in this section was the cost of research and development of drones that the Department of Defense spends every year. This is a tricky subject to deal with as it difficult to understand what funds are being used for drone research as this is not always completely explained. In addition, some parts of the budget are classified. In fact, an entire book could be devoted to the costs of drone research and development. There are probably other items that need to be included in this section but the numbers are likely unknown or are not quantifiable.

In summation, this section of this chapter has tried to provide a preliminary assessment of the costs of drone warfare. Approximately fifty billion has been spent by the United States since 2002 concerning the training and development of drone warfare. This number is much cheaper than the costs of traditional warfare at one trillion and 1.7 trillion. Although, admittedly several costs have not been provided, particularly research and development, the author can still safely say that at this point, drone warfare is much cheaper than the two traditional wars that the United States has recently fought in Afghanistan and Iraq. However, as always more research needs to be done.

This topic requires a book length research project, most likely with a chapter for each of the tables included in this section. An additional paper could be written concerning just the costs of drone research and development. The costs of research and development are a large project that requires a lot of time and energy and is frankly, not pursuable in this section.

CONCLUSION

Besides drone warfare, there are several other ways to combat terrorism and defeat terrorists. In his book, *The Counter-Terrorism Puzzle*, Scholar Boaz Ganor identified six goals of offensive action against a terrorist group adversary. These goals are: preventing terrorist attacks, disrupting the organization's activity routine, revenge and punishment, deterrence, demoralizing the organization's activists and supporters, lifting the moral of the nation's populace, and improper considerations (future elections, etc.).[35] The United States' Government has worked to prevent terrorist attacks using organizations such as the CIA and the Department of Homeland Security. The military component of the War on Terror is highly effective. However, the United States is still lacking in several of these components such as disrupting the organization's activity routine and lifting the moral of the nation's populace, in addition to other recommendations from the author, which will be discussed below.

The first suggestion is an increase in the use of intelligence and intelligence agencies. Something similar to Israeli intelligence systems and training is appropriate. American universities and colleges are lacking undergraduate and graduate programs in intelligence and security such as the Defense and Strategic Studies Master's Program and the Intelligence and National Security Studies Master's Program at the University of Texas at El Paso. These programs allow the United States to have a workforce trained in intelligence and security, in addition to having language skills, allowing analysts to have a solid background in intelligence before entering the workforce. The military should rarely, if ever, be used to fight terrorism; fighting terrorism belongs strictly to the domain of the intelligence community.

In addition, drone targeted killing has its benefits when used correctly. The Israelis use targeted killing to kill terrorists that present an immediate threat. The United States has not been as fastidious in selecting its targets. There is a lot of collateral damage in American targeted killing, including innocent women and children. This is not acceptable. The idea of a drone court presented by Brand and Guiora has some value in that there should be some oversight on presidential power concerning targeted killing. There should be a list of predetermined criteria that are weighed, and only then, can a terrorist be targeted after serious deliberation. Qualified judges must be chosen and approved by the Senate.

In June 2011, President Obama published his National Counterterrorism Strategy. In this strategy, the United States' affirms its commitment to four core principles. These principles are as follows: "adhering to U.S. core values, building security partnerships, applying counterterrorism tools and capabilities appropriately, and building a culture of resilience."[36] Core values include respect for civil rights, civil liberties, human rights, and upholding the rule of law. The president of the United States is not and should not be the unitary actor in deciding American foreign policy as is specified in the United States *Constitution*. Targeted killing should be the last resort and should have some kind of judicial procedure.

It would also be helpful for the United States to assist in more economic development in Middle Eastern countries. Contrary to the views of the American media, the Arab Spring was not caused by a spontaneous quest for democratic governance in the Middle East. Instead, revolutions occurred because food became scarce and jobs were hard to find. America should engage in capitalist economic development and free trade with the Middle East to begin to lift people from poverty. Terrorism becomes attractive when resources are scarce and people are angry.

One aspect of drones that continues to be at the center of the drone debate is proliferation. We wonder if other countries will continue to acquire drones, let alone the lethal drones like the Reaper and the Predator. Once the United States has phased out all the Predators, will these retired drones then become part of another country's arsenal? In her book, *Drones What Everyone Needs to Know*, Professor Sarah Kreps argues that it is not necessarily correct that the "inexorable march toward a world of drones"[37] will occur for four reasons. The first of these is that although it may be easy for anyone to acquire drones, the acquisition of armed drones can only occur for advanced developed countries who are able to supply the technology and resources to necessary to care for this advanced weaponry. The second reason is that although the American public may overwhelmingly support the use of drones, citizens in other developed countries such as Germany do not. In this author's opinion, Americans would not be as apt to support targeted killing if they were actually aware of the consequences to civilian population. Public opinion polls find that Americans overwhelmingly support targeted killing.[38] Hence, education is the main reason for the publication of this book. The third reason that the proliferation of drones may not be as widespread is that the fiscal costs of the weaponry may not justify the benefits. The fourth and final reason for a decreased proliferation of drones is the stipulations regarding the sales of drones. The United States has a bureaucratic assembly approval line for the sale of drones by private American companies. Governments alone are able to buy drones through the American Government with the approval of Congress.[39]

In an article published in the *Journal of Strategic Studies*, Associate Professor Javier Jordan[40] looks at the effects of drone strikes in Pakistan on al-Qa'ida internationally. Jordan's case study focuses on the effects of Pakistani drone strikes on the hierarchical structures, qualified human resources, and key material resources. The study finds that al Qaeda's ability to carry out terrorist attacks in the West is being impaired. However, Jordan's article fails to look at the problem of blowback created by American drone strikes. In addition, the article only focuses on drone strikes in Pakistan not the rest of the world

There is much more research that needs to be completed on the extent of the effectiveness of American drone targeted killing in Afghanistan, Iraq, Libya, Pakistan, Somalia, Syria, and Yemen. Since American targeted killing is relatively new and data can be difficult to find, scholars are still studying whether targeted killing actually decreases terrorism. This book focused primarily on the effects of targeted killing and found that American targeted killing leads to human collateral damage, an increase in anti-Americanism, the dispersal of terrorists, and an increase in suicide bombings in terrorist attacks. However, targeted killing is still a substantial counterterrorism weapon but it should only be used with caution, judicial oversight, and as a last resort.

NOTES

1. Matt Delmot, "Drone Encounters: Noor Behram, Omer Fast, and Visual Critiques of Drone Warfare," *American Quarterly* 65 (1) (2013): 193–202.

Omer Fast, *5000 Feet is the Best*, Internet, 30 Minutes, 2011. Accessed October 4, 2015. http://www.gbagency.fr/en/42/Omer-Fast/#!/5-000-Feet-is-the-Best/site_video_listes/88.

2. Mattea Kramer, "Who Will You Vote for as Drone Commander in Chief?" *The Nation*, March 3, 2016, accessed May 6, 2016, http://www.thenation.com/article/who-will-you-vote-for-as-drone-commander-in-chief/.

3. Ray Sanchez, Tim Lister, Mark Bixler, Sean O'Key, Michael Hogenmiller and Mohammed Tawfeeq, "ISIS goes global: 90 attacks in 21 countries have killed nearly 1,400 people," *CNN*, April 29, 2016, accessed May 29, 2016, http://www.cnn.com/2015/12/17/world/mapping-isis-attacks-around-the-world/.

4. Rosa Brooks, "The Magical Thinking of Killing Mullah Mansour," *Foreign Policy*, May 24, 2016, accessed May 25, 2016, http://foreignpolicy.com/2016/05/24/the-magical-thinking-of-killing-mullah-mansour-taliban-al-Qa'ida-obama/.

5. Amanda Sakuma, "The U.S. Is Way Behind Its Goal of Accepting 10,000 Syrian Refugees," *NBC News*, April 7, 2016, accessed May 29, 2016, http://www.nbcnews.com/storyline/syrias-suffering-families/u-s-way-behind-its-goal-accepting-10-000-syrian-n552521.

6. Lori Robertson, "Stretching Facts on Syrian Refugees," *Factcheck.org*, September 15, 2015, accessed May 29, 2016, http://www.factcheck.org/2015/09/stretching-facts-on-syrian-refugees/.

7. Michael Waltz, "'Bring on the Magic' Using Drones in Afghanistan," in *Drone Wars*, eds. Peter L. Bergen and Daniel Rothenberg (Cambridge: Cambridge University Press, 2015), 211–212.

8. Michael Waltz, "'Bring on the Magic' Using Drones in Afghanistan," in *Drone Wars*, eds. Peter L. Bergen and Daniel Rothenberg (Cambridge: Cambridge University Press, 2015).

9. Dan Hampton, *Viper Pilot, A Memoir of Air Combat* (New York: Ascalon, LLC, 2012), 304.

10. Geoff Dyer and Chloe Sorvino, "$1tn cost of longest US war hastens retreat from military intervention," *The Financial Times,* October 14, 2014, accessed January 25, 2016, http://www.ft.com/cms/s/2/14be0e0c-8255-11e4-ace7-00144feabdc0.html#slide0.

11. For a detailed look at the costs of the War in Afghanistan including the long-term costs of veterans'' benefits, please see *The Costs of War Project* at Brown University, http://watson.brown.edu/costsofwar/.

12. Daniel Trotta, "Iraq war costs U.S. more than $2 trillion: study," *Reuters,* March 14, 2013, accessed January 25, 2016, http://www.reuters.com/article/us-iraq-war-anniversary-idUSBRE92D0PG20130314.

13. For a detailed look at the costs of the Second War in Iraq including the long-term costs of veterans,'' benefits, please see *The Costs of War Project* at Brown University, http://watson.brown.edu/costsofwar/.

14. Nick Mottern, "DRONEFAQ," *Knowdrones.com*, 2015, accessed February 23, 2016, http://www.knowdrones.com/FAQ.pdf.

15. There is no citation for this but I have found this evidence while researching my book.

16. Chris Mailey, "Are UAS More Cost Effective than Manned Flights?" *AUVSI,* October 24, 2013, accessed February 7, 2016, http://www.auvsi.org/hamptonroads/blogs/chris-mailey/2013/10/24/are-uas-more-cost-effective-than-manned-flights.

17. Senior Airmen A.K., "RPAs reach 2 million hours," *United States Air Force,* October 23, 2013, accessed February 8, 2016, http://www.af.mil/News/ArticleDisplay/tabid/223/Article/467346/rpas-reach-2-million-hours.aspx.

18. Tatjana Obrazcova, "Analysis: Predator Training Demand Continues in New Mexico," *50SkyShades*, October 5, 2015, accessed January 22, 2016, http://50skyshades.com/news/manufacturer/analysis-predator-training-demand-continues-in-new-mexico.

19. Tatjana Obrazcova, "Analysis: Predator Training Demand Continues in New Mexico," *50SkyShades*, October 5, 2015, accessed January 22, 2016, http://50skyshades.com/news/manufacturer/analysis-predator-training-demand-continues-in-new-mexico.

20. Tatjana Obrazcova, "Analysis: Predator Training Demand Continues in New Mexico," *50SkyShades,* October 5, 2015, accessed January 22, 2016, http://50skyshades.com/news/manufacturer/analysis-predator-training-demand-continues-in-new-mexico.

21. Tatjana Obrazcova, "Analysis: Predator Training Demand Continues in New Mexico," *50SkyShades*, October 5, 2015, accessed January 22, 2016, http://50skyshades.com/news/manufacturer/analysis-predator-training-demand-continues-in-new-mexico.

22. In the author's experience at a Holloman Air Force Base Media Day (September 10, 2015), experienced USAF pilots were quickly trained and were typically in high-ranking trainer positions after they had been trained on UAV.

23. This screen shot was taken during a Power Point presentation at a Holloman Air Force Media Day on September 10, 2015.

24. The author received a similar figure from a Power Point presentation that was given at a Holloman Airforce Media Day on September 10, 2015.

25. Please also see the following for a similar slide. Dan Gettinger, "Training Drone Pilots," *Center for the Study of the Drone*, June 29, 2015, accessed February 8, 2016, http://dronecenter.bard.edu/?s=Training+Drone+Pilots.

26. Lt. Col. Matt J. Martin with Charles W. Sasser, *Predator: The Remote-Control Air War over Iraq and Afghanistan: A Pilot's Story* (Minneapolis: Zenith Press, 2010), 26.

27. Office of the Undersecretary of Defense (Comptroller), "United States Department of Defense Fiscal Year 2016 Budget Request," *Department of Defense,* February 2015, accessed February 9, 2016, http://comptroller.defense.gov/Portals/45/Documents/defbudget/fy2016/FY2016_Budget_Request_Overview_Book.pdf.

Leo Shane III, "2016 Defense Budget Finally Nailed Down by Congress," *Military Times*, December 18, 2015, accessed February 9, 2016, http://www.militarytimes.com/story/military/capitol-hill/2015/12/18/omnibus-passage-partisan-success/77569538/.

28. Dan Gettinger, "Drones in the Defense Budget," *Center for the Study of the Drone*, February 4, 2015, accessed February 8, 2016, http://dronecenter.bard.edu/drones-in-the-defense-budget/.

Dan Gettinger, "Drones in the Defense Budget: Part Two," *Center for the Study of the Drone*, June 8, 2015, accessed February 8, 2016, http://dronecenter.bard.edu/drones-in-the-defense-budget-part-two/.

29. Dan Gettinger, "Drones in the Defense Budget," *Center for the Study of the Drone*, February 4, 2015, accessed February 8, 2016, http://dronecenter.bard.edu/drones-in-the-defense-budget/.

Dan Gettinger, "Drones in the Defense Budget: Part Two," *Center for the Study of the Drone*, June 8, 2015, accessed February 8, 2016, http://dronecenter.bard.edu/drones-in-the-defense-budget-part-two/.

30. Dan Gettinger, "Drones in the Defense Budget," *Center for the Study of the Drone*, February 4, 2015, accessed February 8, 2016, http://dronecenter.bard.edu/drones-in-the-defense-budget/.

Dan Gettinger, "Drones in the Defense Budget: Part Two," *Center for the Study of the Drone*, June 8, 2015, accessed February 8, 2016, http://dronecenter.bard.edu/drones-in-the-defense-budget-part-two/.

31. Dan Gettinger, "Drones in the Defense Budget," *Center for the Study of the Drone*, February 4, 2015, accessed February 8, 2016, http://dronecenter.bard.edu/drones-in-the-defense-budget/.

Dan Gettinger, "Drones in the Defense Budget: Part Two," *Center for the Study of the Drone*, June 8, 2015, accessed February 8, 2016, http://dronecenter.bard.edu/drones-in-the-defense-budget-part-two/.

32. Dan Gettinger, "Drones in the Defense Budget," *Center for the Study of the Drone*, February 4, 2015, accessed February 8, 2016, http://dronecenter.bard.edu/drones-in-the-defense-budget/.

Dan Gettinger, "Drones in the Defense Budget: Part Two," *Center for the Study of the Drone*, June 8, 2015, accessed February 8, 2016, http://dronecenter.bard.edu/drones-in-the-defense-budget-part-two/.

33. Dan Gettinger, "Drones in the Defense Budget," *Center for the Study of the Drone*, February 4, 2015, accessed February 8, 2016, http://dronecenter.bard.edu/drones-in-the-defense-budget/.

Dan Gettinger, "Drones in the Defense Budget: Part Two," *Center for the Study of the Drone*, June 8, 2015, accessed February 8, 2016, http://dronecenter.bard.edu/drones-in-the-defense-budget-part-two/.

34. Drone Crash Database, *Drone Wars UK*, 2015, accessed February 6, 2016, http://dronewars.net/drone-crash-database/.

35. Boaz Ganor, *The Counter-Terrorism Puzzle* (New Brunswick: Transaction Publishers, 2005).

36. National Strategy for Counterterrorism, June 29, 2011, accessed June 2, 2014, http://www.whitehouse.gov/blog/2011/06/29/national-strategy-counterterrorism.

37. Sarah Kreps, *Drones What Everyone Needs to Know* (Oxford: Oxford University Press, 2016), 63.

38. Pew Research Center, "Public Continues to Support U.S. Drone Strikes," May 28, 2015, accessed May 30, 2016, http://www.people-press.org/2015/05/28/public-continues-to-back-u-s-drone-attacks/.

39. Sarah Kreps, *Drones What Everyone Needs to Know* (Oxford: Oxford University Press, 2016), Chapter 3.

40. Javier Jordan, "The Effectiveness of the Drone Campaign against al Qaeda Central: A Case Study," *Journal of Strategic Studies* 27 (1) (2014).

Bibliography

Abdalla, Amb and Arezou Hassanzadeh. "Al-Qaida and Terrorism in the Arab East: Rise Decline, and the Effects of Doctrine Revisions and the Arab Revolutions." In *Al-Qaida After Ten Years of War A Global Perspective of Successes, Failures, and Prospects* edited by Norman Cigar and Stephanie E. Kramer, 100–127. Quantico, Virginia: Marine Corps University Press, 2011.

Abdul-Ahad, Ghaith. "Al-Qaida's Wretched Utopia and the Battle for Hearts and Minds." *World News*. April 30, 2012. Accessed January 14, 2016. http://www.theguardian.com/world/2012/apr/30/alqaida-yemen-jihadis-sharia-law.

Abrahams, Fred and Sidney Kwiram. "Unacknowledged Deaths." *Human Rights Watch*. May 13, 2012. Accessed May 3, 2016. https://www.hrw.org/report/2012/05/13/unacknowledged-deaths/civilian-casualties-natos-air-campaign-libya.

Ackerman, Spencer and Noah Shachtman. "Almost 1 in 3 U.S. War Planes in a Robot." *Wired*. January 9, 2012. Accessed February 4, 2016, http://www.wired.com/2012/01/drone-report/.

AFP. "Zardari calls for an end to drone strikes." *The Express Tribune*. September 15, 2012. Accessed December 2, 2015. http://tribune.com.pk/story/437202/pakistan-sees-us-as-major-development-partner-pm-ashraf/.

AFP Washington. "ISIS driving up fighter numbers in Iraq, Syria: CIA." *AFP*. September 12, 2014. Accessed March 31, 2016. http://english.alarabiya.net/en/News/middle-east/2014/09/12/CIA-ISIS-has-20–000-to-31–500-fighters-in-Iraq-and-Syria.html.

Ahmad, Mahmoud. "The United States Use of Drones in Pakistan: A Politico-Strategic Analysis." *Asian Affairs: An American Review* 41(1), 2014: 21–29.

Aid, Matthew. "How Many Unmanned Drones is the US Flying?" *Matthewaid.com*. August 20, 2012. Accessed February 5, 2016, http://www.matthewaid.com/post/29822072373/how-many-unmanned-drones-is-the-us-flying.

Airwars. "Cause for Concern." August 2015, accessed April 17, 2016. http://airwars.org/wp-content/uploads/2015/08/airwars-cause-for-concern-civilians-killed-by-coalition.pdf.

Akhtar, Nasreen. "Pakistan, Afghanistan, and the Taliban." *International Journal on World Peace* XXV (4), 2008: 49–73.

Al Arabiya News. "Guidebook on Muslim Spies Reveals al-Qaeda's Fear." *Al Arabiya News*. July 10, 2009. Accessed December 12, 2015. http://www.alarabiya.net/articles/2009/07/10/78352.html.

Ali, Rafid Fadhil. "AQAP's Man in the South: Nasir al-Wuhayshi." In *The Battle for Yemen*, edited by Ramsey Mardini, 156–159. Washington D.C.: The Jamestown Foundation, 2010.

Ali, Imtiaz and Massoud Ansari. "Pakistan Fury as CIA Airstrike on Village Kills 18." *Telegraph*. January 15, 2006. Accessed September 24, 2015. http://www.telegraph.co.uk/news/worldnews/asia/pakistan/1507895/Pakistan-fury-as-CIA-air-strike-on-village-kills-18.html.

Armstrong, Karen. "Wahhabism to ISIS: How Saudi Arabia Exported the Main Source of Global Terrorism." *The New Statesman*. November 27, 2014. Accessed May 20, 2016, http://www.newstatesman.com/world-affairs/2014/11/wahhabism-isis-how-saudi-arabia-exported-main-source-global-terrorism.

Aslam, Wali. "Terrorist Relocation and the Societal Consequences of US Drone Strikes in Pakistan." *The Remote Control Project*. June 2014. Accessed October 10, 2014, http://remotecontrolproject.org/wp-content/uploads/2014/09/Wali-Report.pdf.

Axe, David. *War Bots, U.S. Military Robots Are Transforming War in Iraq, Afghanistan, and the Future*. Ann Arbor, MI: Nimble Books LLC, 2008.

Baker, Peter. "In Terror Shift, Obama Took a Long Path." *The New York Times*. May 27, 2013. Accessed August 21, 2015. http://www.nytimes.com/2013/05/28/us/politics/in-terror-shift-obama-took-a-long-path.html?_r=0.

Bakier, Abdul Hameed. "Al Qaeda in Yemen Supports Southern Secession." In *The Battle for Yemen*, edited by Ramsey Mardini, 35–37. Washington D.C.: The Jamestown Foundation, 2010.

Bartu, Peter. "The Corridor of Uncertainty." In *The Libyan Revolution and Its Aftermath,* edited by Peter Cole and Brian McQuinn, 31–54. Oxford: Oxford University Press, 2015.

BBC News. "Yemen Captures al-Qaeda Members." *BBC News*. March 4, 2004. Accessed June 1, 2015. http://news.bbc.co.uk/2/hi/middle_east/3531657.stm.

BBC News. "Libya IS head 'killed in US air strike." *BBC News*. November 15, 2015. Accessed April 20, 2016. http://www.bbc.com/news/world-africa-34823466.

Benjamin, Medea. *Drone Warfare Killing by Remote Control*. New York: Verso, 2013.

Bergen, Peter L. *The Longest War, The Enduring Conflict between America and al-Qaeda*. New York: Free Press, 2011.

Bergen, Peter and Katherine Tiedemann, Editors. *Talibanistan, Negotiating the Borders between Terror, Politics, and Religion.* Oxford: Oxford University Press, 2013.

Bergen, Peter L and Jennifer Rowland. "Decade of the Drone." In *Drone Wars*, edited by Peter L. Bergen and Daniel Rothenberg, 12–41. Cambridge: Cambridge University Press, 2015.

Boone, Jon (Islamabad) and Peter Beaumont. "Pervez Musharraf Admits Permitting 'A few' US Drone Strikes in Pakistan." *The Guardian*. April 12, 2013. Accessed April 2, 2016. http://www.theguardian.com/world/2013/apr/12/musharraf-admits-permitting-drone-strikes.

Boone, Jon. "US limits Pakistan drone strikes amid political battle over military moves." *The Guardian*. February 5, 2014. Accessed February 5, 2014, http://www.theguardian.com/world/2014/feb/05/us-limits-pakistan-drone-strikes.

Boone, Jon. "US troops' killing of Hamid Karzai's cousin brings claim of 'deep conspiracy." *The Guardian*. March 10, 2011. Accessed August 4, 2015. http://www.theguardian.com/world/2011/mar/10/hamid-karzai-cousin-nato-death-conspiracy-afghanistan.

Botelho, Greg and Barbara Starr. "49 killed in U.S. airstrike targeting terrorists in Libya." *CNN*. February 20, 2016. Accessed April 25, 2016. http://www.cnn.com/2016/02/20/africa/libya-us-airstrike-isis/.

Boucek, Christopher. "Yemen: Avoiding a Downward Spiral." In *Yemen on the Brink*, edited by Christopher Boucek and Marina Ottaway, 1–27. Washington D.C.: Carnegie Endowment for International Peace, 2010.

Boyle, Michael. "The Costs and Consequences of Drone Warfare." *International Affairs*, 89 I, 2013.

Brand, Jeffrey S. and Amos Guiora. "Establishment of a Drone Court: A Necessary Restraint on Executive Power." In *The Legitimacy of Drones: UCAVs for Cross-Border Counterterrorism* edited by S Barela. Burlington, VT: Ashgate, 2015.

Brooks, Rosa. "The Magical Thinking of Killing Mullah Mansour." *Foreign Policy*. May 24, 2016. Accessed May 25, 2016, http://foreignpolicy.com/2016/05/24/the-magical-thinking-of-killing-mullah-mansour-taliban-al-qaeda-obama/.

Brown, Vahid and Don Rassler. *Fountainhead of Jihad, the Haqqani Nexus, 1973–2012*. New York: Columbia University Press, 2013.

Bryden, Matt. "No Quick Fixes: Coming to Terms with Terrorism, Islam, and Statelessness in Somalia." *The Journal of Conflict Studies* (Fall 2003): 24–56.

The Bureau of Investigative Journalism. "Naming the Dead." 2015. Accessed September 16, 2014, http://www.thebureauinvestigates.com/namingthedead/the-dead/?lang=en.

The Bureau of Investigative Journalism. "US Strikes in Yemen, 2002-present." 2015. Accessed April 1, 2016. https://docs.google.com/spreadsheets/d/1lb1hEYJ_omI8lSe33izwS2a2lbiygs0hTp2Al_Kz5KQ/edit.

The Bureau of Investigative Journalism. "Get the Data: Drone Wars Pakistan." 2016. Accessed February 23, 2016, https://www.thebureauinvestigates.com/category/projects/drones/drones-graphs/.

The Bureau of Investigative Journalism. "Get the Data: Drone Wars Somalia." *The Bureau of Investigative Journalism*. 2016. Accessed February 23, 2016, https://www.thebureauinvestigates.com/category/projects/drones/drones-graphs/.

The Bureau of Investigative Journalism. "Get the Data: Drone Wars Yemen." 2016. Accessed February 23, 2016, https://www.thebureauinvestigates.com/category/projects/drones/drones-graphs/.

The Bureau of Investigative Journalism. "Somalia: Minimum and Maximum People Killed by US Drone Strikes, 2007-present." 2016. Accessed May 15, 2016. https://docs.google.com/spreadsheets/d/1-LT5TVBMy1Rj2WH30xQG9nqr8-RXFVvzJE_47NlpeSY/edit#gid=1531042213.

The Bureau of Investigative Journalism. "US Strikes in Somalia: 2007-Present." 2016. Accessed May 15, 2016. https://docs.google.com/spreadsheets/d/1-LT5TVBMy1Rj2WH30xQG9nqr8-RXFVvzJE_47NlpeSY/edit#gid=859698683.

Bush, George W. "President Bush Addresses the Nation." *The Washington Post*. September 20, 2001. Accessed August 26, 2015, http://www.washingtonpost.com/wp-srv/nation/specials/attacked/transcripts/bushaddress_092001.html.

Byman, Daniel. "Why Drones Work." *Foreign Affairs* 92 (32) (2013).

Byman, Daniel. *Al Qaeda, The Islamic State, and The Global Jihadist Movement, What Everyone Needs to Know*. Oxford: Oxford University Press, 2015.

Carter, Chelsea J., Tom Cohen, and Barbara Starr. "U.S. Jet Fighters, Drone Strikes, ISIS Fighters, Convoys in Iraq." *CNN*. August 9, 2014. Accessed February 2, 2016. http://www.cnn.com/2014/08/08/world/iraq-options/.

Center for Civilians in Conflict. "Caring for Their Own: A Stronger Afghani Response to Civilian Harm." 2013. Accessed August 13, 2015. http://civiliansinconflict.org/uploads/files/publications/Afghan_Report_2013_smaller_final.pdf.

The Center for the Study of Targeted Killing. 2016. Accessed September 16, 2014, http://www.targetedkilling.org/.

Chairman, Joint Chiefs of Staff. "No Strike and the Collateral Damage Estimation Methodology." Chairman of the Joint Chiefs of Staff Instruction. February 13, 2009. Accessed August 21, 2015. https://www.aclu.org/files/dronefoia/dod/drone_dod_3160_01.pdf.

Chase, Simon and Ralph Pezzullo. *Zero Footprint*. New York: Mulholland Books, 2016.

Chicago Project on Security and Terrorism. The University of Chicago. 2015. http://cpostdata.uchicago.edu/search_results_new.php.

Chivers, C.J. and Eric Schmitt. "In Strikes on Libya by NATO, an Unspoken Civilian Toll." *The New York Times*. December 17, 2011. Accessed April 22, 2016. http://www.nytimes.com/2011/12/18/world/africa/scores-of-unintended-casualties-in-nato-war-in-libya.html?pagewanted=all&_r=2&.

Chivvis, Christopher S. *Toppling Qaddafi: Libya and the Limits of Liberal Intervention*. Cambridge: Cambridge University Press, 2013.

CNN Wire Staff. "Timeline: Osama bin Laden, Over the Years." *CNN*. May 2, 2011. Accessed December 30, 2015. http://www.cnn.com/2011/WORLD/asiapcf/05/02/bin.laden.timeline/.

Cockburn, Patrick. "War with Isis: Islamic militants have army of 200,000, claims senior Kurdish leader." *Independent*. November 15, 2014. Accessed March 31, 2016. http://www.independent.co.uk/news/world/middle-east/war-with-isis-islamic-militants-have-army-of-200000-claims-kurdish-leader-9863418.html.

Coglan, Tom. "The Taliban in Helmand." In *Decoding the New Taliban, Insights from the Afghan Field* edited by Antonio Giustozzi, 119–153. New York: Columbia University Press, 2010.

Cole, Chris. "Drones in Iraq and Syria: What we know and what we don't." *Drone Wars UK*. July 11, 2014. Accessed March 27, 2016. http://dronewars.net/2014/11/07/drones-in-iraq-and-syria-what-we-know-and-what-we-dont/.

Cole, Chris, Mary Dobbing, and Amy Hailwood. "Convenient Killing Armed Drones and the 'Playstation' Mentality." *The Fellowship of Reconciliation*. September 2010. Accessed September 30, 2014, http://dronewarsuk.files.wordpress.com/2010/10/conv-killing-final.pdf.

Coll, Steve. *Ghost Wars: The Secret History of the CIA, Afghanistan, and bin Laden, from the Soviet Invasion to September 10, 2001*. New York: Penguin Books, 2005.

Columbia Law School Human Rights Clinic. "The Civilian Impact of Drones: Unexamined Costs, Unanswered Questions." *Center for Civilians in Conflict*. 2012. Accessed November 17, 2015. www.civiliansinconflict.org.

Committee on Foreign Relations. "Al Qaeda in Yemen and Somalia: A Ticking Time Bomb." A Report to the Committee on Foreign Relations United States Senate. January 21, 2010. Accessed December 29, 2015. http://www.foreign.senate.gov/imo/media/doc/Yemen.pdf.

Counterterrorism Guide. "Al-Shabaab." *National Counterterrorism Center*. 2016. Accessed May 20, 2016, https://www.nctc.gov/site/groups/al_shabaab.html.

Crichton, Kyle, Gina Lamb, and Rogene Fisher Jacquette. "Timeline of Major Events in the Iraq War." *The New York Times*. 2003. Accessed January 25, 2016. http://www.nytimes.com/interactive/2010/08/31/world/middleeast/20100831-Iraq-Timeline.html?_r=0#/.

Cronin, Audrey Kurth. "Why Drones Fail, When Tactics Drive Strategy." *Foreign Affairs* 92 (44) (2013): 44–54.

Deed, Mary-Jane. "The Arab Spring: Libya's Second Revolution." In *The Arab Spring Change and Resistance in the Middle East*, edited by Mark L. Hass and David W. Lesch, 64–78. Boulder: Westview Press, 2013.

Delmot, Matt. "Drone Encounters: Noor Behram, Omer Fast, and Visual Critiques of Drone Warfare." *American Quarterly* 65 (1) (2013): 193–202.

Department of Defense. "Operation Inherent Resolve Targeted Operations Against ISIL Terrorists." March 29, 2016. Accessed April 1, 2016. http://www.defense.gov/News/Special-Reports/0814_Inherent-Resolve.

Dill, Janina. *Legitimate Targets? Social Construction, International Law and US Bombing*. Cambridge: Cambridge University Press, 2015.

Dobbins, James F. *After the Taliban*. Washington D.C.: Potomac Books, Inc., 2008.

Draper, Lucy. "Report Accuses Israel of Human Rights Abuses in Gaza During Summer War." *Newsweek*. January 22, 2015, accessed August 4, 2016, http://www.newsweek.com/report-accuses-israel-human-rights-abuses-during-summer-war-301275.

Drone Crash Database. "Drone Wars UK." March 31, 2014. Accessed October 15, 2014, http://dronewars.net/drone-crash-database/.

Drones Team. "Somalia: reported US covert actions 2001–2016." *The Bureau of Investigative Journalism*. February 22, 2012. Accessed May 20, 2016, https://www.thebureauinvestigates.com/2012/02/22/get-the-data-somalias-hidden-war/#SOM023.

DW. "US targets al-Shabab leader in Somalia with drone strike." April 2, 2016. Accessed May 19, 2016, http://www.dw.com/en/us-targets-al-shabab-leader-in-somalia-with-drone-strike/a-18232950.

Dyer, Geoff and Chloe Sorvino. "$1tn cost of longest US war hastens retreat from military intervention." *The Financial Times.* October 14, 2014. Accessed January 25, 2016, http://www.ft.com/cms/s/2/14be0e0c-8255–11e4-ace7–00144feabdc0.html#slide0.

Editor's Choice Libya. "Ayesha Qaffadi as a new leader of resistance against NATO and the Libyan terrorists." *South Front.* January 2, 2016. Accessed April 28, 2016, https://southfront.org/ayesha-Qaffadi-as-a-new-leader-of-resistance-against-nato-and-the-libyan-terrorists/.

Entous, Adam, Siobhan Gorman, and Evan Perez. "U.S. Unease Over Drone Strikes." *The Wall Street Journal.* September 26, 2012. Accessed October 12, 2014, http://online.wsj.com/news/articles/SB10000872396390444100404577641520858011452.

Evans, Ryan. "From Iraq to Yemen: Al-Qa'ida's Shifting Strategies." *Combating Terrorism Center at Westpoint.*" October 1, 2010. Accessed January 12, 2016. https://www.ctc.usma.edu/posts/from-iraq-to-yemen-al-qaida%E2%80%99s-shifting-strategies.

Fahim, Kareem. "Houthi Rebels Kill 45 U.A.E. Soldiers in Yemen Fighting." *The New York Times.* September 4, 2015. Accessed December 30, 2015. http://www.nytimes.com/2015/09/05/world/middleeast/27-soldiers-in-saudi-led-force-reported-killed-in-yemen.html?_r=0.

Fair, C. Christine, Karl Kaltenthaler and William Miller. "Pakistani Political Communication and Public Opinion on U.S. Drone Attacks." *Journal of Strategic Studies* 38 (6), 2015.

Faizi, Aimal. "Drones kill any chance of peace in Afghanistan." *Aljazeera.* June 29, 2015. Accessed August 13, 2015. http://www.aljazeera.com/indepth/opinion/2015/06/drones-kill-chance-peace-afghanistan-150628132016117.html.

Fast, Omer. *5000 Feet is the Best.* Internet. 30 Minutes. 2011. Accessed October 4, 2015. http://www.gbagency.fr/en/42/Omer-Fast/#!/5–000-Feet-is-the-Best/site_video_listes/88.

Federal Research Division, Library of Congress. "Al Qaeda in Libya: A Profile." August 2012. Accessed April 25, 2016. https://fas.org/irp/world/para/aq-libya-loc.pdf.

Fergusson, James. *Taliban, The True Story of the World's Most Feared Guerilla Fighters.* London: Transworld Publishers, 2010.

Fergusson, James. *The World's Most Dangerous Place.* Philadelphia: Da Capo Press, 2013.

Ford, Dana and Steve Almasy. "ISIS confirms death of 'Jihadi John'." *CNN.* January 20, 2016. Accessed April 16, 2016. http://www.cnn.com/2016/01/19/middleeast/jihadi-john-dead/.

Foust, Joshua. "Understanding the Strategic and Tactical Considerations of Drone Strikes." *American Security Project.* January 2013. Accessed September 25, 2014 www.americansecurityproject.org 10.

Franco, Claudio. "The Tehrik-E Taliban Pakistan." In *Decoding the New Taliban, Insights from the Afghan Field* edited by Antonio Giustozzi, 247–29. New York: Columbia University Press, 2010.

Friends Committee on National Legislation. "Understanding Drones." *Friends Committee on National Legislation.* 2016. Accessed February 5, 2016, http://fcnl.org/issues/foreign_policy/understanding_drones/.

Qaffadi, Mauammar. *My Vision*. Translated by Angela Parfitt. London: John Blake, ltd, 2005.

Galliot, Jai. *Military Robots, Mapping the Moral Landscape*. Burlington, Vermont: Ashgate Publishing Company, 2015.

Gambhir, Harleen. "ISIS Regional Campaign, December 2015." *Institute for the Study of War*. January 11, 2016. Accessed April 12, 2016. http://www.understandingwar.org/backgrounder/isiss-regional-campaign-december-2015.

Ganor, Boaz. *The Counter-Terrorism Puzzle*. New Brunswick: Transaction Publishers, 2005.

Gardener, Lloyd C. *Killing Machine, The American Presidency in the Age of Drone Warfare*. New York: The New Press, 2013.

Gartenstein-Ross, Daveed. "Success and Failures of the U.S. and NATO Intervention in Libya." *Foundation for the Defense of Democracies*. May 1, 2014. Accessed January 10, 2016. http://www.defenddemocracy.org/media-hit/gartenstein-ross-daveed-successes-and-failures-of-the-us-and-nato-intervention-in-libya/.

Gartenstein-Ross, Daveed. "How Many Fighters Does the Islamic State Really Have?" *War on the Rocks*. February 9, 2015. Accessed March 31, 2015. http://warontherocks.com/2015/02/how-many-fighters-does-the-islamic-state-really-have/.

Gettinger, Dan. "Drones in the Defense Budget." *Center for the Study of the Drone*. February 4, 2015. Accessed February 8, 2016, http://dronecenter.bard.edu/drones-in-the-defense-budget/.

Gettinger, Dan. "Drones in the Defense Budget: Part Two." *Center for the Study of the Drone*. June 8, 2015. Accessed February 8, 2016, http://dronecenter.bard.edu/drones-in-the-defense-budget-part-two/.

Gettinger, Dan. "Training Drone Pilots." *Center for the Study of the Drone*. June 29, 2015. Accessed February 8, 2016, http://dronecenter.bard.edu/?s=Training+Drone+Pilots.

Gentilini, Fernando. *Afghan Lessons, Culture, Diplomacy, and Counterinsurgency*. Translated by Angela Arnone. Washington D.C.: Brookings Institution Press, 2013.

Ghobari, Mohammed. "Yemeni Parliament in Non-binding Vote against Drone Strikes." *Reuters*. December 15, 2013. Accessed January 5, 2016. http://www.reuters.com/article/us-yemen-drones-idUSBRE9BE0EN20131215.

Ghosh, Palash. "I Would Order Air Force To Shoot Down US Drones: Pakistan Opposition Leader Imran Khan." *International Business Times*. October 5, 2012. Accessed January 27, 2015, http://www.ibtimes.com/i-would-order-air-force-shoot-down-us-drones-pakistan-opposition-leader-imran-khan-842017.

Gopal, Anand. "The Taliban in Kandahar." In *Talibanistan*, edited by Peter Bergen and Katherine Tiedemann, 1–68. New York: Oxford University Press, 2013.

The Guardian. "The ISIS papers: leaked documents show how Isis is building its state." December 7, 2015. Accessed April 12, 2016. http://www.theguardian.com/world/2015/dec/07/leaked-isis-document-reveals-plan-building-state-syria.

Guiora, Amos. "Executive Power and Terrorism." Panel presented at Wofford College, Spartanburg, SC March 28, 2016.

Hahn, Peter L. *Missions Accomplished? The United States and Iraq since World War I?* New York: Oxford University Press, 2012.

Hampton, Dan. *Viper Pilot, A Memoir of Air Combat*. New York: Ascalon, LLC, 2012.

Hansen, Stig Jarle. *Al-Shabaab in Somalia.* London: Hurst and Company, 2016.

Harris, Alistair. "Exploiting Grievances: Al-Qaeda in the Arabian Peninsula." In *Yemen on the Brink*, edited by Christopher Boucek and Marina Ottaway. Washington D.C.: Carnegie Endowment for International Peace, 2010.

Hellmich, Christina. *Al-Qaeda from Global Network to Local Franchise.* Black Point, Nova Scotia: Fernwood Publishing ltd, 2011.

The History Channel. "9–11 Attacks." 2016. Accessed March 18, 2016. http://www.history.com/topics/9–11-attacks.

Holewinski, Sarah. "Just Trust Us, The Need to Know More about the Civilian Impact of US Drone Strikes." In *Drone Wars*, edited by Peter L. Bergen and Daniel Rothenberg, 42–70. Cambridge: Cambridge University Press, 2015.

House Hearing 111 Congress. "Effective Counterinsurgency: The Future of the U.S.-Pakistan Military Partnership." Hearing before the full committee of the Committee on Armed Services. April 23, 2009. Accessed October 15, 2014, http://www.gpo.gov/fdsys/pkg/CHRG-111hhrg52666/content-detail.html.

Hudson, Leila, Colin S. Owens, and Matt Flannels. "Drone Warfare: Blowback from the New American Way of War." *Middle East Policy Council*, XVIII, no.3 (Fall 2011).

Human Rights Watch. *Harsh War, Harsh Peace, Abuses by al-Shabaab, the Transitional Federal Government, and AMISOM.* New York: Human Rights Watch, 2010.

Ibrahim, Mohamed. "Somalia and Global Terrorism: A Growing Connection." *Journal of Contemporary African Studies*, 28:3 (2010): 283–295.

International Crisis Group. "Drones: Myths and Realities in Pakistan." May 21, 2013. Accessed September 28, 2015. http://www.crisisgroup.org/en/regions/asia/south-asia/pakistan/247-drones-myths-and-reality-in-pakistan.aspx.

Iraq Body Count. 2016. Accessed March 27, 2016. https://www.iraqbodycount.org/.

Irwin, Lewis G. *Disjointed Ways, Disunified Means: Learning From America's Struggle to Build an Afghan Nation.* Carlisle, PA: Strategic Studies Institute, 2012.

Islamic Emirate of Afghanistan. "Statement of Islamic Emirate regarding prolonging stay of American invaders in Afghanistan." October 16, 2015. Accessed November 11, 2015. http://shahamat-english.com/statement-of-islamic-emirate-regarding-prolonging-stay-of-american-invaders-in-afghanistan/.

al Jaber, Faisal bin ali. "What is the value of an innocent life, Mr. Obama: $100K in a plastic bag?" *CNN Opinion.* June 12, 2015. Accessed January 14, 2016. http://www.cnn.com/2015/06/12/opinions/drone-victims-relatives-sue-us-opinion/.

Jaeger, David A. and Zahra Siddique. "Are Drone Strikes Effective in Afghanistan and Pakistan? On the Dynamics of Violence between the United States and the Taliban." *Institute for the Study of Labor.* December 2011. Accessed September 25, 2014. http://ftp.iza.org/dp6262.pdf.

Al Jazeera and Agencies. "Islamic State 'has 50,000 fighters in Syria'." *al-Jazeera.* August 19, 2014. Accessed March 31, 2016. http://www.aljazeera.com/news/middleeast/2014/08/islamic-state-50000-fighters-syria-2014819184258421392.html.

Al Jazeera and Agencies. "ISIL training camp' in Libya targeted by US." *al-Jazeera.* February 20, 2016. Accessed April 20, 2016. http://www.aljazeera.com/news/2016/02/deaths-reported-raids-isil-camp-libya-160219131122223.html.

Johnsen, Gregory D. "Yemen's Passive Role in the War on Terrorism." In *The Battle for Yemen*, edited Ramsey Mardini, 87–90. Washington D.C.: The Jamestown Foundation, 2010.

Johnsen, Gregory D. "Did an 8-Year-Old Spy for America?" *The Atlan*tic. September 2013. Accessed January 13, 2016. http://www.theatlantic.com/magazine/archive/2013/09/did-an-8-year-old-spy-for-america/309429/.

Johnsen, Gregory D. *The Last Refuge*. New York: W.W. Norton & Company, 2013.

Johnston, Patrick B. and Anoop K. Sarbahi. "The Impact of US Drone Strikes on Terrorism in Pakistan and Afghanistan." *Princeton University*. July 14, 2013. Accessed September 30, 2014, https://webspace.princeton.edu/users/esocweb/ESOC%20website%20publications/JS_DronesTerrorism.pdf.

Jordan, Javier. "The Effectiveness of the Drone Campaign against al Qaeda Central: A Case Study." *Journal of Strategic Studies* 27 (1) (2014): 4–29.

Kalin, Stephen. "U.S. says it killed IS militant who killed Marine in Iraq." *Reuters*. April 4, 2016. Accessed April 8, 2016. http://www.reuters.com/article/us-mideast-crisis-iraq-usa-idUSKCN0X00JS.

Kamat, Anjali and Ahmad Shokr. "Libya." In *Dispatches from the Arab Spring Understanding the New Middle East*, edited by Paul Amar and Vijay Prashad, 157–203. Minneapolis: University of Minnesota Press, 2013.

Key Data. "Predator RQ-1 / MQ-1 / MQ-9 Reaper UAV, United States of America." *Air-force-technology.com*. 2016. Accessed February 4, 2016, http://www.airforce-technology.com/projects/predator-uav/.

Khan, Azmat. "Understanding Yemen's Al Qaeda Threat." *Frontline*. May 29, 2012. Accessed January 14, 2016. http://www.pbs.org/wgbh/frontline/article/understanding-yemens-al-qaeda-threat/.

Khan, Ismail. "Night Raid Kills Nek, Four Other Militants: Wana Operation." *Dawn*. June 19, 2004. Accessed November 19, 2015. http://www.dawn.com/2004/06/19/top1.htm.

Kilcullen, David. *Blood Year, Islamic State and the Failures of the War on Terror*. London: Hurst and Company, 2016.

Kramer, Mattea. "Who Will You Vote for as Drone Commander in Chief?" *The Nation*. March 3, 2016. Accessed May 6, 2016, http://www.thenation.com/article/who-will-you-vote-for-as-drone-commander-in-chief/.

Kreps, Sarah. *Drones What Everyone Needs to Know*. Oxford: Oxford University Press, 2016.

Lewis, Lawrence. "Drone Strikes: Civilian Casualty Considerations Executive Summary." *Joint and Coalition Operational Analysis*. Accessed September 16, 2014. http://cna.org/sites/default/files/research/Drone_Strikes.pdf (This source has since been removed from the Internet).

Lieven, Anatol. *Pakistan, A Hard Country*. New York: Penguin Books, 2011.

Lister, Charles. "Profiling the Islamic State." *Brookings Doha Center Analysis Paper 12*. November 2014. Accessed March 24, 2016. http://www.brookings.edu/~/media/Research/Files/Reports/2014/11/profiling%20islamic%20state%20lister/en_web_lister.pdf.

Lobban, Richard A. Jr. and Christopher H. Dalton. *Libya, History and Revolution*. Santa Barbara, CA: Praeger, 2014.

Mailey, Chris. "Are UAS More Cost Effective than Manned Flights?" *AUVSI*. October 24, 2013. Accessed February 7, 2016, http://www.auvsi.org/hamptonroads/blogs/chris-mailey/2013/10/24/are-uas-more-cost-effective-than-manned-flights.

Mantzikos, Ioannis. "Somalia and Yemen: The Links between Terrorism and State Failure." *Digest of Middle East Studies* 20:2 (2011): 242–260.

Marcus, Jonathan. "Iraq and Syria opinion poll - the world's most dangerous survey?" *BBC*. September 9, 2015. Accessed May 30, 2016, http://www.bbc.com/news/world-middle-east-34173549.

Marsden, Peter. *The Taliban, War and Religion in Afghanistan*. New York: Zed Books Ltd., 2002.

Martin, Lt. Col. Matt J. with Charles W. Sasser. *Predator The Remote-Control Air War over Iraq and Afghanistan: A Pilot's Story*. Minneapolis: Zenith Press, 2010.

Martinez, Luis. "Number of ISIS Fighters in Iraq and Syria Drops, Increases in Libya, US Official Says." *ABC News*. February 4, 2016. Accessed March 24, 2016. http://abcnews.go.com/International/number-isis-fighters-iraq-syria-drops-increases-libya/story?id=36715635.

Maruf, Harun and Jeff Seldin. "Shootings in Somalia's Capital Leave 7 Dead." *Voice of America*. April 2, 2016. Accessed May 23, 2016, http://www.voanews.com/content/us-drone-attack-kills-top-al-shaba-assassin-somalia/3265379.html.

Masood, Salman. "C.I.A. Leaves Base in Pakistan Used for Drone Strikes. *The New York Times*. December 1, 2011. Accessed September 24, 2015, http://www.nytimes.com/2011/12/12/world/asia/cia-leaves-pakistan-base-used-for-drone-strikes.html?_r=0.

Masters, Jonathan. "Targeted Killings," *Council on Foreign Relations*. May 23, 2013. Accessed July 14, 2014, http://www.cfr.org/counterterrorism/targeted-killings/p9627.

Mazetti, Mark and Scott Shane. "As New Drone Policy Is Weighed, Few Practical Effects Are Seen." *The New York Times*. March 21, 2013. Accessed August 2015. http://www.nytimes.com/2013/03/22/us/influential-ex-aide-to-obama-voices-concern-on-drone-strikes.html.

McCants, William. *The ISIS Apocalypse, The History, Strategy, and Doomsday Vision of the Islamic State*. New York: St. Martin's Press, 2015.

McLeary, Paul with Adam Rawnsley. "White House Drone Report." *The Situation Report, The National Security Daily Brief from Foreign Policy*. July 1, 2016. Accessed August 4, 2016. http://link.foreignpolicy.com/view/54da7bb010defb923d8b49ec47b0k.1gd6/2052ebeb.

Meleagrou-Hitchens, Alexander and Abdihakim Ainte. "The Return of al-Shabaab." *The Daily Beast*. March 15, 2016. Accessed on May 25, 2016, http://www.thedailybeast.com/articles/2016/03/15/the-return-of-al-shabaab.html.

Melzer, Nils. *Targeted Killing in International Law*. Oxford: Oxford University Press, 2008.

Miller, Greg. "CIA seeks new authority to expand Yemen drone campaign." *The Washington Post*. April 18, 2012. Accessed December 30, 2015. https://www.washingtonpost.com/world/national-security/cia-seeks-new-authority-to-expand-yemen-drone-campaign/2012/04/18/gIQAsaumRT_story.html.

Mohammed, Riyadh. "Islamic State Expands: Up to 100,000 People Have Joined, Experts Say." *Mashable*. August 26, 2014. Accessed March 31, 2016. http://mashable.com/2014/08/26/100000-people-join-islamic-state/#aJF7hmnZ9sqN.

Monitoring Group on Somalia and Eritrea. "Report on the Monitoring Group on Somalia and Eritrea pursuant to the Security Council Resolution 2002 (2011)." *United Nations Security Council*. 2011. Accessed May 20, 2016, http://www.un.org/ga/search/view_doc.asp?symbol=S/2012/544.

Monthana, Ibrahim. "How Drones Help al-Qa'ida,." *The New York Times*. June 13, 2012. Accessed September 23, 2014. http://www.nytimes.com/2012/06/14/opinion/how-drones-help-al-qaeda.html?_r=0.

al-Moshki, Ali Ibrahim. "Three US Drone Strikes in Eight Days." *Yemen Times*. February 4, 2015. Accessed December 30, 2015. http://www.yementimes.com/en/1857/news/4860/Three-US-drone-strikes-in-eight-days.htm.

Mottern, Nick. "DRONEFAQ." *Knowdrones.com*. 2015. Accessed February 23, 2016, http://www.knowdrones.com/FAQ.pdf.

Mullen, Jethro. "Al Qaeda's second in command killed in Yemen strike; successor named." *CNN*. June 16, 2015. Accessed January 12, 2016. http://www.cnn.com/2015/06/16/middleeast/yemen-aqap-leader-killed/.

Mullick, Haider Ali Hussein. "Power by Proxy: Al-Qaida in Pakistan." In *Al-Qaida After Ten Years of War A Global Perspective of Successes, Failures, and Prospects* edited by Norman Cigar and Stephanie E. Kramer, Pages 160–179. Quantico, Virginia: Marine Corps University Press, 2011.

Murphy, Martin N. *Somalia: The New Barbary?* New York: Columbia University Press, 2011.

Musharraf, Pervez. *In the Line of Fire, A Memoir*. New York: Free Press, 2006.

Mutiga, Murithi. "Refugees urge Kenyan leaders to rethink closure of Dadaab camp." *The Guardian*. May 13, 2016. Accessed May 13, 2016. http://www.theguardian.com/global-development/2016/may/13/refugees-urge-kenyan-leaders-to-rethink-closure-of-dadaab-camp.

Nance, Malcolm. *Defeating ISIS, Who They Are, How They Fight, What they Believe*. New York: Skyhorse Publishing, 2016.

North Atlantic Treaty Organization. "NATO and Libya (Archived)." *NATO*. November 9, 2015. Accessed April 25, 2016. http://www.nato.int/cps/en/natolive/topics_71652.htm.

National Consortium for the Study of Terrorism and Responses to Terrorism. "Global Terrorism Database, Somalia." 2015. Accessed May 17, 2016. https://www.start.umd.edu/gtd/search/Results.aspx?chart=overtime&casualties_type=&casualties_max=&country=182.

National Strategy for Counterterrorism. June 29, 2011. Accessed June 2, 2014, http://www.whitehouse.gov/blog/2011/06/29/national-strategy-counterterrorism.

Nawa, Fariba. "Women and the Drug Trade in Afghanistan." In *Under the Drones*, edited by Shahzad Bashir and Robert D. Crews, 236–256. Cambridge: Harvard University Press, 2012.

Nojumi, Neamatollah, Dyan Maurana, and Elizabeth Stites. *After the Taliban, Life and Security in Rural Afghanistan*. Lanham: Rowman & Littlefield Publishers, Inc., 2009.

Northrop Grumman. "Firescout Unprecedented Persistent Situational Awareness." 2016. Accessed February 5, 2016, http://www.northropgrumman.com/Capabilities/firescout/Pages/default.aspx.

Northrop Grumman. "Global hawk." 2016. Accessed February 5, 2016, http://www.northropgrumman.com/capabilities/globalhawk/Pages/default.aspx.

Obrazcova, Tatjana. "Analysis: Predator Training Demand Continues in New Mexico." *50SkyShades*. October 5, 2015. Accessed January 22, 2016, http://50skyshades.com/news/manufacturer/analysis-predator-training-demand-continues-in-new-mexico.

Office of the Undersecretary of Defense (Comptroller). "United States Department of Defense Fiscal Year 2016 Budget Request." *Department of Defense.* February 2015. Accessed February 9, 2016, http://comptroller.defense.gov/Portals/45/Documents/defbudget/fy2016/FY2016_Budget_Request_Overview_Book.pdf.

Open Canada Staff. "Gregory Johnsen on Yemen, the U.S., and Drones." *Open Canada.org*. December 14, 2012. Accessed January 5, 2016. https://www.opencanada.org/features/gregory-johnsen-on-yemen-the-u-s-and-drones.

The Other Jihadis. "The Return of al-Shabaab." *The Daily Beast.* March 15, 2016. Accessed May 25, 2016, http://www.thedailybeast.com/articles/2016/03/15/the-return-of-al-shabaab.html.

Pakistan Body Count.org. "Pakistan Body Count." 2015. Accessed September 16, 2015, http://pakistanbodycount.org/suicide_bombing.

Peritz, Aki and Eric Rosenbach. *Find, Fix, Finish, Inside the Counterterrorism Campaigns That Killed Bin Laden and Devastated Al-Qaeda.* New York: Public Affairs, 2012.

Peters, Gretchen S. "The Taliban and the Opium Trade." In *Decoding the New Taliban, Insights from the Afghan Field* edited by Antonio Giustozzi, 7–22. New York: Columbia University Press, 2010.

Pew Research Center. "Pakistani Public Opinion Ever More Critical of U.S." June 27, 2012. Accessed January 27, 2015, http://www.pewglobal.org/2012/06/27/pakistani-public-opinion-ever-more-critical-of-u-s/.

Pew Research Center. "Public Continues to Support U.S. Drone Strikes." May 28, 2015. Accessed May 30, 2016, http://www.people-press.org/2015/05/28/public-continues-to-back-u-s-drone-attacks/.

Pew Research Global Attitudes Project. "U.S. Image in Pakistan Falls No Further Following bin Laden Killing Support for Campaign against Extremists Wanes." June 21, 2011. Accessed October 12, 2014, http://www.pewglobal.org/2011/06/21/u-s-image-in-pakistan-falls-no-further-following-bin-laden-killing/.

Pew Research Global Attitudes Project. "Global Indicators Database: Drone Strikes." 2014. Accessed January 27, 2015, http://www.pewglobal.org/database/indicator/52/survey/16/.

Pew Research Global Attitudes & Trends. "A Less Gloomy Mood in Pakistan." August 27, 2014. Accessed May 5, 2016, http://www.pewglobal.org/2014/08/27/a-less-gloomy-mood-in-pakistan/.

Pianan, Eric. "U.S. Shoots Itself In the Foot By Accidentally Arming ISIS." *The Fiscal Times*. June 4, 2015. Accessed April 1, 2016. http://www.thefiscaltimes.com/2015/06/04/Fog-War-US-Has-Armed-ISIS.

Plaw, Avery. *Targeting Terrorists A License to Kill?* Cornwall, England: Ashgate, 2008.

Plaw, Avery and Matthew S. Fricker. "Tracking the Predators: Evaluating the US Drone Campaign in Pakistan." *International Studies Perspectives*, 13 (2012): 344–365.

Pope, Laurence. *The Demilitarization of American Diplomacy*. New York: Palgrave Macmillan, 2014.

Povoledo, Elisabetta. "Giovanni Lo Porto Was Known as Aid Worker Drawn to Needy." *New York Times*. April 23, 2015. http://www.nytimes.com/2015/04/24/world/europe/american-drone-strike-italian-aid-worker-giovanni-lo-porto.html.

Priest, Dana. "Surveillance Operation in Pakistan Located and Killed Al Qaeda Official." *Washington Post*. May 15, 2005. Accessed November 19, 2015. http://www.washingtonpost.com/wp-dyn/content/article/2005/05/14/AR2005051401121.html.

Rae, James DeShaw. *Analyzing the Drone Debates: Targeted Killing, Remote Warfare, and Military Technology*. New York: Palgrave Macmillan, 2014.

Raghavan, Sudarsan. "In Yemen, U.S. airstrikes breed anger, and sympathy for al-Qaeda." *The Washington Post*. May 29, 2012. Accessed October 9, 2014. http://www.washingtonpost.com/world/middle_east/in-yemen-us-airstrikes-breed-anger-and-sympathy-for-al-qaeda/2012/05/29/gJQAUmKI0U_story_2.html.

Raghavan, Sudarsan. "When U.S. drones kill civilians, Yemen's government tries to conceal it." *The Washington Post*. December 24, 2012. Accessed January 13, 2016. https://www.washingtonpost.com/world/middle_east/when-us-drones-kill-civilians-yemens-government-tries-to-conceal-it/2012/12/24/bd4d7ac2–486d-11e2–8af9–9b50cb4605a7_story.html.

Rashid, Ahmed. *Pakistan on the Brink*. New York: Penguin, 2012.

Rasmussen, Anders Fogh. "Monthly Bress Briefing." *North Atlantic Treaty Organization*. November 3, 2011. Accessed April 22, 2016. http://www.nato.int/cps/en/natolive/opinions_80247.htm.

Rawlence, Ben. *City of Thorns, Nine Lives in the World's Largest Refugee Camp*. New York: Picador, 2014.

Reilly, Mollie. "Obama Told Aides He's 'Really Good At Killing People,' New Book 'Double Down' Claims." *The Huffington Post*. November 4, 2013. Accessed August 17, 2016, http://www.huffingtonpost.com/2013/11/03/obama-drones-double-down_n_4208815.html.

Reuters. "Abandoned Al Qaeda notebook found in Yemen features blueprint for creating Islamic state." *Rueters*. July 8, 2014. Accessed October 10, 2014. http://www.nydailynews.com/news/national/abandoned-al-qaeda-notebook-features-blueprint-creating-islamic-state-article-1.185853477.

Riviera, Denise. "Harmony and Mayhem in Somalia." *The Advocate*. May 6, 2016. Accessed May 6, 2016, http://gcadvocate.com/2016/05/06/harmony-mayhem-somalia/.

Robertson, Lori. "Stretching Facts on Syrian Refugees." *Factcheck.org* September 15, 2015. Accessed May 29, 2016, http://www.factcheck.org/2015/09/stretching-facts-on-syrian-refugees/.

Rodriguez, Alex. *Los Angeles Times*. "Pakistani Death Squads go after Informants to U.S. Drone Program." December 28, 2011. Accessed August 21, 2014, http://articles.latimes.com/2011/dec/28/world/la-fg-pakistan-death-squads-20111228.

Rogers, Simon. "Drones by country: who has all the UAVs?" *The Guardian*. August 3, 2012. Accessed February 4, 2016, http://www.theguardian.com/news/datablog/2012/aug/03/drone-stocks-by-country.

Roggio, Bill. "'Good Taliban' leader Mullah Nazir killed in US drone strike." *The Long War Journal*. January 3, 2013. Accessed December 2, 2015. http://www.longwarjournal.org/archives/2013/01/good_taliban_leader_1.php.

Roggio, Bill. "US Strikes in Pakistan, HVTs." *The Long War Journal*. 2015. Accessed November 17, 2015. http://www.longwarjournal.org/pakistan-strikes-hvts.

Roggio, Bill. "Islamic State's governor for Aleppo reportedly killed in airstrike." *Threat Matrix, A Blog of the Long War Journal*. March 4, 2016. Accessed April 16, 2016. http://www.longwarjournal.org/archives/2016/03/islamic-states-governor-for-aleppo-reportedly-killed-in-airstrike.php.

Rohde, David. "7 Months, 10 Days in Captivity." *The New York Times*. October 18, 2009. Accessed September 14, 2015. http://www.nytimes.com/2009/10/18/world/asia/18hostage.html?pagewanted=all&_r=0.

Rohde, David. "My Guards Absolutely Feared Drones." In *Drone Wars*, edited by Peter L. Bergen and Daniel Rothenberg, 9–11. Cambridge: Cambridge University Press, 2015.

Ross, Alice. "Who is dying in Afghanistan's 1000 -plus drone strikes?" *The Bureau of Investigative Journalism*. July 24, 2014. Accessed August 4, 2015. https://www.thebureauinvestigates.com/2014/07/24/who-is-dying-in-afghanistans-1000-plus-drone-strikes/.

Rubin, Alissa J. "U.S. Report on Afghan War Finds Few Gains in 6 Months." *The New York Times*. April 29, 2010. Accessed September 16, 2015. http://www.nytimes.com/2010/04/30/world/asia/30afghan.html?_r=0.

Rubin, Barnett R. *Afghanistan from the Cold War to the War on Terror*. New York: Oxford University Press, 2013.

Rubin, Barnett and Abubakar Siddique. "Resolving the Pakistan-Afghanistan Stalemate." In *Afghanistan from the Cold War through the War on Terror*, edited by Barnett Rubin. Oxford: Oxford University Press, 2013.

Rutig, Thomas. "Loya Paktia's Insurgency (I) The Haqqani Network As An Autonomous Entity." In *Decoding the New Taliban, Insights from the Afghan Field* edited by Antonio Giustozzi, 57–100. New York: Columbia University Press, 2010.

Sakuma, Amanda. "The U.S. Is Way Behind Its Goal of Accepting 10,000 Syrian Refugees." *NBC News*. April 7, 2016. Accessed May 29, 2016, http://www.nbcnews.com/storyline/syrias-suffering-families/u-s-way-behind-its-goal-accepting-10–000-syrian-n552521.

Sanaa. "Newsmaker: Yemen's Saleh dance "on heads of snakes" nears end." *Reuters*. November 23, 2011. Accessed January 11, 2015. http://www.reuters.com/article/us-yemen-saleh-idUSTRE7AM1UR20111123.

Sanchez, Ray, Tim Lister, Mark Bixler, Sean O'Key, Michael Hogenmiller and Mohammed Tawfeeq. "ISIS goes global: 90 attacks in 21 countries have killed nearly 1,400 people." *CNN*. April 29, 2016. Accessed May 29, 2016, http://www.cnn.com/2015/12/17/world/mapping-isis-attacks-around-the-world/.

Scahill, Jeremy. "The Secret US War in Pakistan." *The Nation*. November 23, 2009. Accessed November 19, 2015. http://www.thenation.com/article/secret-us-war-pakistan/.

Scahill, Jeremy. *Dirty Wars, The World is a Battlefield.* New York: Nation Books, 2013.

Scahill, Jeremy. "The CIA's Secret Sites in Somalia." *The Nation.* December 10, 2014. Accessed May 20, 2016, http://www.thenation.com/article/cias-secret-sites-somalia/.

Scahill, Jeremy, editor. *The Assassination Complex.* New York: Simon & Schuster, 2016.

Schmidt, Michael S. and Mark Mazzeti. "A Top ISIS Leader Is Killed in an Airstrike, the Pentagon Says." *The New York Times.* March 25, 2016. Accessed April 8, 2016. http://www.nytimes.com/2016/03/26/world/middleeast/abd-al-rahman-mustafa-al-qaduli-isis-reported-killed-in-syria.html?_r=0.

Schmitt, Eric and Michael S. Schmidt. "Omar the Chechen, a Senior Leader in ISIS, Dies After U.S. Airstrike." *The New York Times.* March 15, 2016. Accessed March 8, 2016. http://www.nytimes.com/2016/03/15/world/middleeast/omar-chechen-isis-killed-us-airstrike-syria.html.

Semple, Michael. "Afghanistan: The First Theater of War." In *Assessing the War on Terror* edited by Mohammed Ayob and Etga Ugur, 47–70. Boulder: Lynne Rienner Publishers, 2013.

Senior Airmen A.K. "RPAs reach 2 million hours." *United States Air Force.* October 23, 2013. Accessed February 8, 2016, http://www.af.mil/News/ArticleDisplay/tabid/223/Article/467346/rpas-reach-2-million-hours.aspx.

Serle, Jack. "Obama 2015 Pakistan Drone Strikes." *The Bureau of Investigative Journalism.* January 5, 2015. Accessed September 16, 2015. https://www.thebureauinvestigates.com/2015/01/05/obama-2015-pakistan-drone-strikes/.

Serle, Jack and Payenda Sargand. "Get the data: A list of US air and drone strikes, Afghanistan 2015." *The Bureau of Investigative Journalism.* February 12, 2015. Accessed August 13, 2015. https://www.thebureauinvestigates.com/2015/02/12/us-drone-war-afghanistan-list-american-air-strikes-2015/#AFG066.

Shachtman, Noah. "Military Stats Reveal Epicenter of US Drone War." *Wired.com.* November 9, 2012. Accessed February 23, 2016, http://www.wired.com/2012/11/drones-afghan-air-war/.

Shachtman, Noah and Spencer Ackerman. "Let's Admit It: The U.S. is at War in Yemen Too." *Wired.* June 14, 2012. Accessed December 30, 2015. http://www.wired.com/2012/06/yemen-war/.

Shaffer, Anthony. *Operation Dark Heart, Spycraft and Special Ops on the Frontlines of Afghanistan-And the Path to Victory.* New York: Thomas Dunne Books, 2010.

Shah, Saeed. "Anti-Americanism rises in Pakistan over U.S. motives." McClatchy D.C. September 7, 2009. Accessed January 27, 2015, http://www.mcclatchydc.com/2009/09/07/74966/anti-americanism-rises-in-pakistan.html.

Shah, Saeed, Adam Entous, and Gordon Lubold. "U.S. Threatens to Withhold Pakistan Aid." *The Wall Street Journal.* August 21, 2015. Accessed October 5, 2015. http://www.wsj.com/articles/u-s-threatens-to-withhold-pakistan-aid-1440163925.

Shalal, Andrea and Susan Heavey. "Islamic State Second in Command, Likely Killed: U.S.." *Reuters.* March 26, 2016. Accessed April 8, 2016. http://www.reuters.com/article/us-mideast-crisis-syria-isis-idUSKCN0WR12V.

Shane, Leo III. "2016 Defense Budget Finally Nailed Down by Congress." *Military Times.* December 18, 2015. Accessed February 9, 2016, http://www.militarytimes.com/story/military/capitol-hill/2015/12/18/omnibus-passage-partisan-success/77569538/.

Shane, Scott. "Yemen's Leader Praises U.S. Drone Strikes." *The New York Times*. September 29, 2012. Accessed December 20, 2015. http://www.nytimes.com/2012/09/29/world/middleeast/yemens-leader-president-hadi-praises-us-drone-strikes.html.

Shane, Scott. "Families of Drone Strike Victims in Yemen File Suit in Washington." *The New York Times*. June 8, 2015. Accessed January 14, 2016. http://www.nytimes.com/2015/06/09/world/middleeast/families-of-drone-strike-victims-in-yemen-file-suit-in-washington.html?_r=0.

Shay, Shaul. *Somalia between Jihad and Restoration*. New Brunswick: Transaction Publishers, 2008.

Sheva, Arutz. "Drone strike kills another top ISIS commander." *Israel National News*. March 31, 2106. Accessed Aril 8, 2016. http://www.israelnationalnews.com/News/News.aspx/210136#.Vwhodpvmphc.

al-Shishani, Murad Batal. "Ibrahim al-Rubaish: New Religious Ideologue of al-Qaeda in Saudi Arabia Calls for Revival of Assassination Tactic." In *The Battle for Yemen*, edited by Ramsey Mardini, 137–139. Washington D.C.: The Jamestown Foundation, 2010.

al-Shishani, Murad Batal. "From Yemen to Detroit: The Expanding Influence of AQAP's Sa'id al Shihri." In *The Battle for Yemen*, edited by Ramsey Mardini 143–147. Washington D.C.: The Jamestown Foundation, 2010.

al-Shishani, Murad Batal. "Adel al-Abbab: Al-Qaeda in the Arabian Peninsula's Religious Ideologue." In *The Battle for Yemen*, edited by Ramsey Mardini 151–155. Washington D.C.: The Jamestown Foundation, 2010.

al-Shishani, Murad Batal. "An Assessment of the Anatomy of al-Qaeda in Yemen: Ideological and Social Factors." *Terrorism Monitor*, VIII (9) (March 5, 2010). Accessed December 29, 2015. http://www.jamestown.org/programs/tm/single/?tx_ttnews%5Btt_news%5D=36111&tx_ttnews%5BbackPid%5D=457&no_cache=1#.VoLWDi_Slhc.

Singer, P.W. *Wired for War: The Robotics Revolution and Conflict in the Twenty-first Century*. New York: Penguin Press, 2009.

Sixta, Christine Marie. "The Illusive Third Wave: Are Female Terrorists the New "New Women" In Developing Societies?" *The Journal of Women, Politics, and Policy* 29 (2), 2008.

Sixta Rinehart, Christine. "Sixta Rinehart: Iran all Over Again." T*he State* (Columbia, SC). February 10, 2011. Editorial-opinion Extra.

Sixta Rinehart, Christine. *Volatile Social Movements and the Origins of Terrorism: The Radicalization of Change*. Lanham, MD: Lexington Books, 2012.

Skaine, Rosemarie. *Women of Afghanistan in the Post-Taliban Era, How Lives Have Changed and Where They Stand Today*. Jefferson, North Carolina: McFarland & Company, Inc., Publishers, 2008.

Solomon, Hussein. *Terrorism and Counter-terrorism in Africa*. New York: Palgrave Macmillan, 2015.

Sorenson, David S. *An Introduction to the Modern Middle East,* 2nd ed. Boulder: Westview Press, 2014.

Spencer Ackerman and agencies. "US drone strike kills key al-Shabaab leader in Somalia" *The Guardian*. April 1, 2016. Accessed May 19, 2016, http://www.theguardian.com/world/2016/apr/01/us-military-somalia-airstrike-al-shabaab-terrorist-targets.

Spencer, Richard. "How Yemen's US-backed ex-dictator is tearing his country apart." *The Telegraph*. March 28, 2015. Accessed December 30, 2015. http://www.telegraph.co.uk/news/worldnews/middleeast/yemen/11502027/How-Yemens-US-backed-ex-dictator-is-tearing-his-country-apart.html.

Sputnik International. "US Drone Strikes in Pakistan Breed Terrorism - President to Sputnik." March 16, 2015. Accessed October 5, 2015. http://sputniknews.com/asia/20150316/1019558238.html.

St. John, Ronald Bruce. *Libya from Colony to Revolution*. Oxford: One World, 2012.

Stanford International Human Rights & Conflict Resolution Clinic. "Drones: An Overview." 2012. Accessed October 12, 2014, http://www.livingunderdrones.org/background-and-context/.

Starr, Barbara. "U.S. officials say 6,000 ISIS fighters killed in battles." *CNN*. January 22, 2015. Accessed March 31, 2016. http://www.cnn.com/2015/01/22/politics/us-officials-say-6000-isis-fighters-killed-in-battles/.

Stenersen, Anne. "The Relationship between al-Qaeda and the Taliban." In *Talibanistan*, edited by Peter Bergen and Katherine Tiedemann, 69–93. New York: Oxford University Press, 2013.

Stewart, Phil. "U.S. strikes al Shabaab training camp in Somalia, more than 150 killed." *Reuters*. March 8, 2016. Accessed May 19, 2016, http://www.reuters.com/article/us-usa-somalia-dronestrike-idUSKCN0W91XW.

Strick Van Linschoten, Alex and Felix Kuehn. *An Enemy We Created, The Myth of the Taliban-Al Qaeda Merger in Afghanistan*. Oxford: Oxford University Press, 2012.

al-Sumait, Fahed, Nele Lenze, and Michael C. Hudson. *The Arab Uprisings, Catalysts, Dynamics, and Trajectories*. Lanham: Rowman & Littlefield, 2015.

Swanson, Will and Ariel Zirulnick. "Hiking the 'Rat Tracks,' Hunting for al-Shabaab." *Foreign Policy*. May 20, 2016. Accessed May 20, 2016, http://foreignpolicy.com/2016/05/20/hiking-the-rat-tracks-hunting-for-al-shabab/?utm_source=Sailthru&utm_medium=email&utm_campaign=New%20Campaign&utm_term=%2AEditors%20Picks.

Swift, Christopher. "The Drone Blowback Fallacy Strikes in Yemen Aren't Pushing People to Al Qaeda." *Foreign Affairs*. July 1, 2012. Accessed October 12, 2014. http://www.foreignaffairs.com/articles/137760/christopher-swift/the-drone-blowback-fallacy.

Swift, Christopher. "The Boundaries of War? Assessing the Impact of Drone Strikes in Yemen." In *Drone Wars*, edited by Peter L. Bergen and Daniel Rothenberg, 71–88. Cambridge: Cambridge University Press, 2015.

Syrian Network for Human Rights. "The Main Conflict Parties Who are Killing Civilians in Syria." November 19, 2015. Accessed April 12, 2016. http://sn4hr.org/wp-content/pdf/english/68_civilians_killed_in_new_attacks_by_international_coalition_forces_en.pdf.

Syrian Network for Human Rights. "New Attacks by the International Coalition Forces Result in Losses and Civilian Victims." February 13, 2016. Accessed April 12, 2016. http://sn4hr.org/wp-content/pdf/english/68_civilians_killed_in_new_attacks_by_international_coalition_forces_en.pdf.

Taarnby, Michael. "Yemen's Committee for Dialogue: Can Jihadists Return to Society?" In *The Battle for Yemen*, edited by Ramsey Mardini, 78–82. Washington D.C.: The Jamestown Foundation, 2010.

Taj, Farhat. "Drone Attacks-A Survey." *International News*. March 5, 2009. Accessed January 28, 2015, http://www.thenews.com.pk/TodaysPrintDetail.aspx?ID=165781&Cat=9&dt=3/4/2009.

Taylor, Lauren. "Yemen's Counter-terrorism Unit." *Al-Jazeera*. September 17, 2008. Accessed January 13, 2015. http://www.aljazeera.com/news/middleeast/2007/07/2008525185044972283.html.

Tayler, Letta. "A Wedding that Became a Funeral." *Human Rights Watch*. February 19, 2014. Accessed January 5, 2016. https://www.hrw.org/report/2014/02/19/wedding-became-funeral/us-drone-attack-marriage-procession-yemen.

Telegraph. "Al-Qaeda Number three 'Killed by CIA Spy Plane' in Pakistan." *Telegraph*. December 4, 2005. Accessed November 19, 2015. http://www.telegraph.co.uk/news/worldnews/asia/pakistan/1504718/Al-Qaeda-number-three-killed-by-CIA-spy-plane-in-Pakistan.html.

Terrill, W. Andrew. *The Conflict in Yemen and U.S. National Security*. Carlisle, PA: Strategic Studies Institute, 2011.

The Asia Foundation. "Afghanistan in 2014: A Survey of the Afghan People." November 18, 2014. Accessed November 2, 2015. http://asiafoundation.org/resources/pdfs/Afghanistanin2014final.pdf.

Thompson, Mark. "Costly Flight Hours." *Time*. April 2, 2013. Accessed February 5, 2016, http://nation.time.com/2013/04/02/costly-flight-hours/.

Transcript 6. "Top Secret America." *PBS Frontline*. 2010. Accessed January 5, 2016. http://www.pbs.org/wgbh/frontline/film/topsecretamerica.

Trotta, Daniel. "Iraq war costs U.S. more than $2 trillion: study." *Reuters*. March 14, 2013. Accessed January 25, 2016. http://www.reuters.com/article/us-iraq-war-anniversary-idUSBRE92D0PG20130314.

United Nations Assistance Mission in Afghanistan (UNAMA). "Afghanistan Midyear Report 2014, Protection of Civilians in Armed Conflict." July 14, 2015. Accessed August 3, 2015. http://unama.unmissions.org/LinkClick.aspx?fileticket=m_XyrUQDKZg%3d&tabid=12254&mid=15756&language=en-US.

United States Department of Justice. "White Paper." *Department of Justice*. February 4, 2013. Accessed March 29, 2016, http://msnbcmedia.msn.com/i/msnbc/sections/news/020413_DOJ_White_Paper.pdf.

United States Department of Justice. "Attorney General Eric Holder Speaks at Northwestern University School of Law." *Department of Justice*. Chicago, IL. March 5, 2012.

United States Navy. "CNIC Camp Lemonnier Djibouti." 2016. Accessed May 18, 2016. http://cnic.navy.mil/regions/cnreurafswa/installations/camp_lemonnier_djibouti.html.

University of Oxford. "University of Oxford: National Public Opinion Poll of Libya." February 14, 2012. Accessed May 30, 2016, http://www.prnewswire.com/news-releases/university-of-oxford-national-public-opinion-poll-of-libya-139308408.html.

Van Bijlert, Martine. "The Taliban in Zabul and Uruzgan." In *Talibanistan*, edited by Peter Bergen and Katherine Tiedemann, 94–127. New York: Oxford University Press, 2013.

Vandewalle, Dirk. *A History of Modern Libya*. Cambridge: Cambridge University Press, 2006.

Waltz, Michael. "'Bring on the Magic' Using Drones in Afghanistan." In *Drone Wars*, edited by Peter L. Bergen and Daniel Rothenberg, 209–214. Cambridge: Cambridge University Press, 2015.

Watson Institute. *The Costs of War Project*. Brown University. 2016. Accessed May 27, 2016, http://watson.brown.edu/costsofwar/.

Weisgerber, Marcus. "Peeling the Onion Back on the Pentagon's Special Operations Budget." *Defense One*. January 27, 2015. Accessed May 26, 2016, http://www.defenseone.com/management/2015/01/peeling-onion-back-pentagons-special-operations-budget/103905/.

Weiss, Michael and Hassan Hassan. *ISIS Inside the Army of Terror*. New York: Regan Arts, 2015.

The White House. "Executive Order." Whitehouse.gov. July 1, 2016. Accessed August 4, 2016. https://www.whitehouse.gov/the-press-office/2016/07/01/executive-order-united-states-policy-pre-and-post-strike-measures.

Whitlock, Craig. "Remote U.S. Base at Core of Secret Operations." *The Washington Post*. October 25, 2012. Accessed May 20, 2016, https://www.washingtonpost.com/world/national-security/remote-us-base-at-core-of-secret-operations/2012/10/25/a26a9392–197a-11e2-bd10–5ff056538b7c_story.html.

Williams, Brian Glyn. "The Cia's Covert Predator Drone War in Pakistan, 2004–2010: The History of An Assassination Campaign." *Studies in Conflict and Terrorism* 33, 2010: 871–892.

Williams, Brian Glyn. *Predators, The CIA's Drone War on al Qaeda*. Dulles, Virginia: Potomac Books, 2013.

Wilner, Alex S. *Deterring Rational Fanatics*. Philadelphia: University of Pennsylvania Press, 2015.

Wittes, Benjamin and Gabriella Blum. *The Future of Violence*. New York: Basic Books, 2015.

Woods, Chris. "Drone strikes in Yemen: The civilian massacre the US neither confirms nor denies." *The Bureau of Investigative Journalism*. March 29, 2012. Accessed January 12, 2015. https://www.thebureauinvestigates.com/2012/03/29/the-civilian-massacre-the-us-will-neither-confirm-nor-deny/.

Woods, Chris. "CIA drone strikes violate Pakistan's sovereignty, says senior diplomat." *The Guardian*. August 3, 2012. Accessed January 27, 2015, http://www.theguardian.com/world/2012/aug/03/cia-drone-strikes-violate-pakistan.

Woods, Chris. *Sudden Justice, America's Secret Drone Wars*. New York: Oxford University Press, 2013.

Woods, Chris. "Iraq." Airwars.org. 2016. Accessed February 23, 2016. http://airwars.org/.

Woods, Chris. "Strikes in Iraq." *Airwars.org*. 2016. Accessed January 1, 2016. http://airwars.org/.

Woods, Chris. "Strikes in Syria." *Airwars.org*. 2016. Accessed January 1, 2016. http://airwars.org/.

Woods, Chris. "Syria." *Airwars.org*. 2016. Accessed February 23, 2016, http://airwars.org/.

Woods, Chris and Alice K. Ross. "Revealed: US and Britain launched 1,200 drone strikes in recent wars." *The Bureau Investigates*. December 4, 2012. Accessed March 27, 2016. https://www.thebureauinvestigates.com/2012/12/04/revealed-us-and-britain-launched-1200-drone-strikes-in-recent-wars/.

Wood, Paul and Richard Hall. "U.S. Killing More Civilians in Iraq, Syria Than it Acknowledges. *USA Today*. February 4, 2016. Accessed April 12, 2016. http://www.usatoday.com/story/news/world/2016/02/02/us-killing-more-civilians-iraq-and-syria-than-acknowledges-globalpost/79686772/.

The World Bank. "Overview Libya." March 31, 2016. Accessed May 5, 2016. http://www.worldbank.org/en/country/libya/overview.

Worth, Robert F. "10 are Killed in Bombings at Embassy in Yemen." *The New York Times*. September 17, 2008. Accessed January 12, 2016. http://www.nytimes.com/2008/09/18/world/middleeast/18yemen.html?_r=0.

Worth, Robert F. Worth, Mark Mazzetti, and Scott Shane. "Drone Strikes' Risks to Get Rare Moment in the Public Eye." *The New York Times*. February 5, 2013. Accessed January 12, 2015. http://www.nytimes.com/2013/02/06/world/middleeast/with-brennan-pick-a-light-on-drone-strikes-hazards.html?_r=0.

Yenne, Bill. *Birds of Prey: Predators, Reapers, and America's Newest UAVs in Combat*. North Branch, MN: Specialty Books, 2010.

al-Zahrani, Fares Shuwail. "To Our People in the South." May 2004. Accessed December 29, 2015. www.ukht-benladn.net/mhadrat_show-446.html.

Zucchino, David. "Afghans describe relatives' deaths in recent U.S. drone strike." *Los Angeles Times*. December 1, 2013. Accessed August 4, 2015. http://articles.latimes.com/2013/dec/01/world/la-fg-afghanistan-drones-20131201.

Zulaika, Joseba. "Drones and Fantasy in US Counterterrorism." *Journal for Cultural Research*, 18(2), 2014: 171–187.

Index

About the Author

Christine Sixta Rinehart is an Assistant Professor of Political Science at the University of South Carolina Palmetto College. Her first book, *Volatile Social Movements and the Origins of Terrorism: The Radicalization of Change* was first published in 2014 in paperback and in 2012 in hardback. Dr. Sixta Rinehart's work focuses on terrorism, counterterrorism, and American Foreign Policy.